FROM COMPETITION AT HOME TO COMPETING ABROAD

FROM COMPETITION AT HOME TO COMPETING ABROAD

A CASE STUDY OF INDIA'S HORTICULTURE

Aaditya Mattoo
Deepak Mishra
Ashish Narain

OXFORD
UNIVERSITY PRESS

THE WORLD BANK

OXFORD
UNIVERSITY PRESS

YMCA Library Building, Jai Singh Road, New Delhi 110 001

Oxford University Press is a department of the University of Oxford. It furthers the University's objective of excellence in research, scholarship, and education by publishing worldwide in

Oxford New York
Auckland Cape Town Dar es Salaam Hong Kong Karachi Kuala Lumpur Madrid
Melbourne Mexico City Nairobi New Delhi Shanghai Taipei Toronto

With offices in
Argentina Austria Brazil Chile Czech Republic France Greece Guatemala Hungary
Italy Japan Poland Portugal Singapore South Korea Switzerland Thailand Turkey Ukraine Vietnam

Oxford is a registered trademark of Oxford University Press
in the UK and in certain other countries.

Published in India
by Oxford University Press, New Delhi

First published 2007

ISBN-13: 978-0-19-568593-0
ISBN-10: 0-19-568593-8

Typeset in AGaramond 11/15
by Eleven Arts, Keshav Puram, Delhi 110 035
Printed in India at De-Unique, New Delhi 110 018
Published by Oxford University Press
YMCA Library Building, Jai Singh Road, New Delhi 110 001

Contents

Acknowledgements vii

Abbreviations and Acronyms ix

Tables, Figures, and Boxes xi

Executive Summary xv

1. An Overview 1
2. International Trade in Horticulture 11
3. Is there a Rationale for Protection? 21
4. What Undermines Comparative Advantage? The Role of Domestic Constraints 29
5. Is the External Environment Conducive to Export Expansion? 55
6. Priorities for Domestic Reform and International Engagement 80

Appendix 1 83

Appendix 2 85

Bibliography 89

Acknowledgements

Shatanjaya Dasgupta and Francis K.T. Ng made a significant contribution to the empirical analysis in this report. Outstanding administrative assistance was provided by Jyoti Sriram. The work was carried out under the guidance of Ijaz Nabi (Sector Manager), Sadiq Ahmed (Sector Director), and Michael Carter (India's Country Director). Deepak Ahluwalia, M. Ataman Aksoy, Tercan Baysan, Guang Zhe Chen, Dina Umali-Deninger, Ejaz Ghani, Stephen Howes, Steven Jaffee, Garry Pursell, Mona Sur, Simon Thomas, John Wilson, and Martien Van Niuewkoop provided helpful comments at various stages of the report.

The authors thanks peer reviewers John Nash (World Bank), Veena Jha (UNCTAD), and Jayanta Roy (CII) for their comments on the report.

The authors are grateful to R. Gopalan and Vandana Agarwal (Ministry of Commerce) for their guidance on the project. The study also benefited greatly from useful discussions held with Ramesh Chand (NCAP), S.C. Garg (NHB), Ashok Gulati (IFPRI), Vijay Sardana (CITA), and Anil Sharma (NCAER). ORG Marg conducted the field survey for the study.

Abbreviations and Acronyms

AEZ	Agricultural Export Zone
AMS	Aggregate Measure of Support
APEDA	Agricultural and Processed Food Products Export Development Authority
APHIS	Animal and Plant Health Inspective Service
APMA	Agriculture Produce and Marketing Act
ATF	Aviation Turbine Fuel
BIS	Bureau of Indian Standards
CCFS	Central Committee of Food Standards
CEPII	Centre d'Etudes Prospectives et d'Informations Internationales
CIF	Cost, Insurance, and Freight
CII	Confederation of Indian Industry
CITA	Centre for International Trade in Agriculture and Agro-based Industries
DGCIS	Directorate General of Commercial Intelligence and Statistics
ECA	Essential Commodities Act
EMS	Equivalent Measure of Support
EPCG	Export Promotion Capital Goods
ERS	Economic Research Service
EU	European Union
FAIDA	Food and Agriculture Integrated Development Action
FAO	Food and Agriculture Organization
FDA	Food and Drug Administration
FDI	Foreign Direct Investment
FIPB	Foreign Investment Promotion Board
FOB	Free on Board
FTA	Free Trade Agreement
GAP	Good Agricultural Practice
GATT	General Agreement on Tariffs and Trade
GoI	Government of India
GSP	Generalized System of Preferences
HACCP	Hazard Analysis and Critical Control Point
HH	Household
HS	Harmonized System
IFPRI	International Food Policy Research Institute
ISO	International Organization for Standardization
JNPT	Jawaharlal Nehru Port Trust
MFN	Most Favoured Nation
MRL	Maximum Residue Levels
MT	Metric Tonne
NABL	National Accreditation Board for Testing and Calibration Laboratories
NAFED	National Agricultural Cooperative Marketing Federation of India
NAFTA	North American Free Trade Agreement
NCAER	National Council of Applied Economic Research

NCAP	National Centre for Agricultural Economics and Policy Research
NHB	National Horticulture Board
NPC	Nominal Protection Coefficient
NQC	National Quality Campaign
NTB	Non Tariff Barriers
NTM	Non Tariff Measures
OECD	Organization for Economic Co-operation and Development
PSE	Producer Subsidy Equivalent
QRs	Quantitative Restrictions
R&D	Research & Development
RCA	Revealed Comparative Advantage
SEZ	Special Economic Zone
SPS	Sanitary and Phyto-sanitary
SSA	Sub Saharan Africa
SSG	Special Safeguards
SSI	Small Scale Industry
STQC	Standardization Testing and Quality Certification
TAMP	Tariff Authority of Major Ports
TB	Tariff Barriers
TFYP	Tenth Five Year Plan
TRQ	Tariff Rate Quota
UNCTAD	United Nations Conference on Trade and Development
URAA	Uruguay Round Agreement on Agriculture
WTO	World Trade Organization

Tables, Figures, and Boxes

TABLES

3.1: India's Applied Tariffs for Main Crops, 1990–2002 21

5.1: Domestic Support by European Communities, Marketing Year 2001–2 56
5.2: Domestic Support by US, Marketing Year 2000 57
5.3: Export Subsidy Expenditures for Horticulture Products 57
5.4: Applies MFN Tariffs for Fresh Fruits and Vegetables in the Quad Countries 58
5.5: Percentage of Tariff Lines at Different Levels/Intervals in Rich Countries: Fresh 58
5.6: Percentage of Tariff Lines at Different Levels/Intervals in Rich Countries: Processed 59
5.7: Percentage of Tariff lines on Fresh Fruits and Vegetables in Selected Developing Countries by Tariff Level 59
5.8: Applied Tariffs in Major Markets of Indian Crops, 2002–3 60
5.9: Seasonal Variation: Current MFN Import Tariffs for Selected Fruits and Vegetables—EC, US, and Japan 61
5.10: EU and US Applied and Specific Tariffs on Selected Indian Fruits and Vgetable, 2004 (in per cent) 64
5.11: Number of Tariff Quotas by Member and Product Group 65
5.12: TRQs on Fruit and Vegetable Imports into the EU 66
5.13: EU's Applied Tariffs on Selected Crops by Selected Country, 2003 67
5.14: US Applied Tariffs on Selected Crops by Selected Country, 2003 67
5.15: Tariff Escalation of Agricultural Products in US and EU, 2003 68
5.16: Detailed Information on the EU's Tariff Rate Quotas 76
5.17: Mangoes: NTMs Faced by a Sample of Indian Firms in Export Markets 78

A1.1: Assumed Potential Competitors 84
A2.1: Commodity-wise production in the States Surveyed 85
A2.2: Final Choice of Commodities, States, and Districts 87

FIGURES

1.1: India's Agriculture Sector 2
1.2: India's Cost of Production and Market Share 4
1.3: Barriers to Trade: External vs Domestic Factors 5
1.4: NPCs for Horticulture: All India, 1997–2002 7
1.5: Price Variation across Space and Seasons for Domestic and Imported Apples and Potatoes 8

2.1: Production and Growth of Horticulture in Selected Countries, 1980–2004 11
2.2: Yield per Hectare in Fruits and Vegetables in Selected Countries, 1990–2004 12
2.3: Decomposition of Growth Rate of Horticulture Sector 13
2.4: Distance has a Considerable Influence on US Imports and India's Exports, but Chile's Products seem to have Broken the Distance Barrier 15
2.5: The Perception of Exporters about the Quality of their Exports 17
2.6: Exports and Imports in Agriculture and Horticulture 17
2.7: Destination of India's Exports, 2003 18
2.8: Exports and Imports in Agriculture and Horticulture 19

3.1: Nominal Protection Coefficients for Fruits and Vegetables: All India, 1997–2002 23
3.2: Regional Price Variations for Fruits and Vegetables, 2002 26
3.3: Imports and Domestic Arrivals in India, Apple and Green Peas, 2002 27

4.1: Evolution of Price/Cost across the Supply Chain 30
4.2: Explaining India's High Delivery Costs 31
4.3: Perishability of Horticulture Products can be a Potential Barrier to Trade 32
4.4: Barriers to Trade in Horticulture 33
4.5: India's International Transport Costs vis-à-vis International Standards 34
4.6: Dwell Time in India vs International Norms (in hours) 35
4.7: An International Comparison of the Performance of India's Ports and its Maritime Infrastructure 38
4.8: International Transport Cost Incurred at Departure and Destination Ports 42
4.9: Wastage at Various Stages of the Value Chain (per cent of total production) 43
4.10: Infrastructure in Local Mandis is Perceived by the Farmers to be Inadequate and of Poor Quality 44
4.11: Considerable Variation in the Amount and Quality of Infrastructure Available in the Local Mandis across States 45
4.12: Secular Decline in Investment in Indian Agriculture 47
4.13: A Typical Supply Chain of Fruits and Vegetables in India vs More Mature Markets (for example, US) 49

5.1a: Entry Price for Lemons and Tomatoes into the EU, 2004–5 62
5.1b: Entry Price for Apples and Grapes into the EU, 2004–5 63

5.2: Impact of, and Response to, Higher Standards in Destination Markets, 2004–5 70
5.3: Signals that Reflect a Failure to Meet Rising Standards and Factors Stimulating a Positive Response, 2004–5 72
5.4: Impact of Rising Standards on Production Costs, 2004–5 74
5.5: Overall Impact of Rising Standards on Exports, 2004–5 75
5.6: Evaluation of Barriers in Four Destination Markets, 2004–5 75

A2.1: Composition of Farmers Surveyed Commodity-wise 85
A2.2: States Surveyed 86

BOXES

4.1: Role of Transportation Technology in Expanding Exports of Perishable Agricultural Products in the US 39
4.2: Ennore—A Successful Model for Port Privatization in India 40
4.3: Regulatory Policies Affecting Domestic Trade in Agriculture 47
4.4: India's National Horticulture Mission 52

5.1: India's Domestic Standards on Agricultural Products 73

Executive Summary

WHY IS POLICY SO DEFENSIVE FOR A SECTOR THAT IS SUPPOSEDLY COMPETITIVE?

There is persuasive evidence that India has a comparative advantage in agriculture. Yet its share in global agriculture exports is miniscule and its domestic market is increasingly protected. In global trade negotiations, India's efforts have been directed less towards the elimination of all distortions at home and abroad, than towards retaining the right to protect. To develop a case and a strategy for both domestic reforms and proactive engagement in international negotiations, it is critical to understand why a large, low-cost producer of agricultural commodities needs such strong protection and has such a feeble presence in the global market.

We address this puzzle in the context of the horticulture sector, one of the most dynamic segments of Indian agriculture and international trade. The sector represents a 'picture of the future' and encompasses the major challenges that Indian agriculture faces in a world where food patterns are changing because of increasing incomes; delivery schedules are more demanding with the emergence of supermarkets in retail; and barriers to trade such as tariffs and subsidies are less important than the challenges posed by high sanitary and phyto-sanitary (SPS) standards. The horticulture sector also represents a miniature 'picture of agriculture' in that here too India is a large, low-cost cost producer but has a tiny share in global trade and seeks to protect its own market.

A complete understanding of India's competitiveness in high value agricultural commodities has proved elusive because the sector has not been subject to an integrated analysis—from farm to retail. Most existing studies have examined only a single stage of the supply chain, for example, productivity at the farm, efficiency of agricultural markets, or problems associated with poor infrastructure, in isolation from the rest of the supply chain. Therefore, policymakers are not in a position to assess how problems identified at a single stage of the supply chain compare and interact with (often unidentified) problems in the rest of the supply chain.

This study undertakes a supply chain analysis of ten horticultural products, based on primary surveys that covered 1400 farmers, 200 commission agents, and 65 exporters across fifteen different Indian states. It adds to the existing literature by laying out the cost and price structure of all agents and all markets in the supply chain, thus providing an integrated view of the sector. The study makes it possible to analyse the relative importance of factors that put Indian produce at a disadvantage, relative to that from other countries, and hence to identify priorities for policy reform.

The study comes to the conclusion that the competitiveness of India's horticultural sector depends critically on efficient logistics and the ability to comply with health, safety, and quality standards abroad. Efficient logistics depend on improved communication, transport and distribution, suggesting that a services revolution in India is desirable not just for its own sake but also for the transformation of India's agricultural trade. At the same time, India must take an aggressive position in the World Trade Organization (WTO) negotiations, seeking not just significantly lower levels of foreign protection, but also much greater transparency, simplicity, and predictability in foreign trade regimes. Its negotiating position would be strengthened by a willingness to reform its own trade regime.

IS INDIAN HORTICULTURE REALLY COMPETITIVE?

In order to assess the competitiveness of Indian horticulture, we estimate the nominal protection coefficient (NPC) under the importable hypothesis, which is the ratio of domestic to international prices (appropriately modified to take into account transport and other costs) for the period 1997–2002 for 10 major horticultural crops. This exercise confirms the findings of a number of previous studies that India's horticulture sector is competitive in terms of prices, with the average NPC consistently below one between 1997 and 2002. The results of this exercise however should be viewed with caution because data limitations preclude effective control for differences in quality and for fine distinctions in varieties between imported and home grown commodities.

While India's horticulture sector as a whole appears to be competitive, the level of competitiveness varies considerably across commodities.[1] Among fruits, bananas appear to be the most competitive: domestic prices on average are consistently less than half the price of imports when they arrive in local markets and there is very little variation over time during the period 1997–2002. However, it should be recognized that a large part of this difference in prices may be explained by the significant difference in quality between domestic and imported bananas. Grapes and oranges (lime/lemon) have slightly higher NPCs, but both appear to be becoming more competitive over time. Apples and mangoes appear less competitive but mostly because there were weather-related shocks in some of the recent years and in the case of mangoes, from the use of cost, insurance, and freight (CIF) price from Pakistan as the reference international price. Apart from potatoes, all vegetables are found to be highly competitive and even potatoes appear to be competitive vis-à-vis imports from more distant markets. Onions and tomatoes are consistently competitive. It is interesting to note that onions remained cheaper than imports even in the crisis year 1998.

THEN WHY ARE DOMESTIC TARIFFS SO HIGH?

It is not obvious why India has such high tariffs for horticulture products that are clearly competitive. We attach greatest credence to a combination of two explanations. First, tariffs for particular products seem to be designed to protect domestic produce in all regions and all seasons. Because of large inefficiencies in the domestic logistics system and storage and marketing infrastructure, there are significant variations in prices across regions (for example, wholesale prices of apples in Chennai can be more than two and half-times those in Delhi) and across seasons (for example, the difference in potato prices in Delhi between peak and off-peak periods can be as much as 350 per cent). Thus, a tariff that does not vary across regions and seasons needs to be consistently high if the objective is to protect domestic produce even in markets far from production regions and peak seasons. The second explanation is that the level of tariffs is not always sensitive to the specific current circumstances of individual products, but is set at roughly similar, sometimes historically determined levels across a wide range of products—a form of policy hysteresis.

India's resultant tariff structure, however, does have the virtue of being less complex, and more uniform and transparent than the tariff structure for fruits and vegetables in, say the European Union (EU). However, the high tariffs, vestiges of a desire for self-sufficiency, and comprehensive protection of domestic producers, clearly penalize consumers in regions located far from production sources and in seasons far from the harvest season. While the net costs of protection are well understood, what is notable here is that tariffs restrict imports even in places and during times when domestic products are not available in markets. The problem is compounded by the fact that India has high costs of transport and storage, not as much in terms of the explicit prices of these services as in the implicit costs imposed by the high rates of product wastage and reduced shelf-life. The result is that tariffs subsidize the inefficient transportation, distribution, and storage system more than they subsidize the actual producer. Therefore, the net welfare loss from protection is even larger because there are few offsetting benefits in terms of rents and revenues for domestic agents.

WHAT IMPEDES INDIA'S EXPORTS?

India's exports face three major impediments. The first is the 'logistical tax' imposed by high delivery costs which significantly erode the production cost advantage enjoyed by Indian farmers. The second is the gap between the high standards required by governments and buyers, especially in richer countries, and the low standards and weak conformity assessment mechanisms in India. The third problem is the trade policy barriers Indian exporters face in foreign markets.

The supply chain analysis indicates that from a purely cost perspective the biggest impediments to the competitiveness of India's horticultural exports lie *outside* the sector rather than *inside* it. The average price at the farm-gate for a typical horticulture product was found to be only 12–15 per cent of the price at which the product is retailed in the destination market, so that a 20 per cent improvement in yields will lead only to a 2.4–3 percentage point reduction in the final price. On the other hand, a 20 per cent reduction in international transportation cost alone will reduce final prices by as much as 8–10 percentage points. However, this is not to understate the importance for Indian farmers of developing better varieties of horticulture products that are internationally more competitive, and of using methods of production that are internationally more acceptable.

Logistical Tax

Three factors explain India's high logistical tax: (a) geography, which is important but not decisive; (b) poor transport and storage infrastructure, as well as policies that have led to the uneven utilization of existing infrastructure, and the slow creation of new infrastructure; and (c) high marketing costs due to the fragmentation of the supply chain, attributable to policies that have inhibited both integration and competition along the supply chain. Inadequate attention to these issues is perhaps the single-most important reason why past prognoses about dramatic growth in India's food sector have not come true.

India's logistics system is significantly less efficient than that in other countries. Based on a number of comparisons, across destinations, markets, modes of transport and commodities traded, we find that on average, India's international transportation costs are 15–30 per cent higher than those in other countries. For example, it costs $790 to transport 1 metric tonne of grapes from India to Netherlands—nearly three times the cost of transporting grapes from Chile to Netherlands, although Chile is twice as far from the Netherlands as India.

Production in horticulture is scale neutral but storing, transport, marketing, and distribution (wholesale and retail) are subject to economies of scale. A number of government policies have prevented the realization of these economies. To increase its control over storage, marketing, and distribution of agricultural commodities during periods of food shortages within the country, the Government enacted the Essential Commodities Act (ECA) (1955), Agricultural Produce and Marketing Act (APMA) (1972), and the Prevention of Black Marketing and Maintenance of Supplies of Essential Commodities Act (1980). These policies have prevented the free mobility of agricultural produce and thus segmented the Indian domestic market into many smaller markets. The government has also imposed restrictions on foreign investment in the retail of agricultural commodities, and on both foreign and domestic private investment in wholesale. These restrictions have collectively discouraged and/or prevented the private sector from undertaking large-scale investment in agricultural storage, marketing, or processing activities—an example of horizontal fragmentation preventing desirable vertical integration. The result is that today there is no large, organized, efficient pan-Indian supply chain in the agricultural sector, including in horticulture.

Some of the policies that have weakened private investment incentives and contributed to India's fragmented, small-scale, and inefficient marketing system are now being changed. The central government and several state governments have lifted measures that restricted private storage and interstate movement of grain and other essential foods. Licenses are no longer needed to establish food-processing firms, and regulations restricting their size have been mostly eliminated. In addition, foreign direct investment in food processing and marketing—with the exception of retail marketing—is now automatically approved for investments involving up to 51 per cent foreign equity. A relaxation of the ban on foreign investment in retailing is also being considered.

Technical Barriers to Trade

Differences between domestic and foreign standards are an important barrier to trade in horticulture but the evidence of their impact on India's exports is mixed. India's fresh exports that are directed towards regional markets with relatively low standards face few SPS measures—described as a pattern of 'defensive commercialization' in a recent World Bank study on the impact of standards on horticultural trade. There is clear evidence that stringent industrial country standards have been a prohibitive barrier for some products such as mangoes. In the case of other products such as grapes, exporters have incurred temporary losses due to the rejection of consignments that do not meet standards. In general, producers who have had to make adjustments in response to higher standards have typically incurred significant consignment-specific or recurrent transactions costs. Between 1995 and 2000, it is estimated that nearly 270 SPS measures were introduced on imports of fruits and vegetables in WTO member countries. Quite apart from the high level of standards, the lack of harmonized technical standards and treatments is itself a hindrance to trade. While some standards in industrial countries serve protectionist goals, it is not easy to distinguish between such standards and those that serve legitimate health and safety goals. Moreover, some of the most demanding standards today are imposed not by foreign governments but by foreign buyers, reflecting increased consumer sensitivity to quality in all its dimensions.

Inadequacies in domestic standard setting can legitimize foreign barriers. Concerns have been expressed about weaknesses in the Indian system of enforcement of pesticide-related regulations and standards of hygiene, which undermine the credibility of domestic certification processes. India today has a plethora of regulatory authorities tackling these issues, often in a piecemeal manner. They include various central ministries (Commerce, Food and Consumer Affairs, and Food Processing), export promotion institutions, standard-setting bodies (as many as 24), a host of regulating bodies at the state level, as well as institutions like the Bureau of Indian Standards (BIS). A major revamp of the food law aimed at consolidating responsibilities and jurisdiction, as well as establishing closer links to international standards, is now underway, although there is no clear time frame for its completion.

Finally, standards need to be seen not only as an impediment but also as an opportunity. Rising standards accentuate underlying supply chain strengths and weaknesses, and thus impact differently on the competitive position of individual countries and different market participants. Some countries and industries are even using high quality and safety standards to successfully position themselves in global markets. The challenge for India is to take a proactive rather than a defensive approach, so that it can exploit its strengths and overcome weaknesses to benefit from the emerging commercial and regulatory context. Some evidence for this more optimistic view is to be found in our survey of exporters: 15 per cent of exporters reported that rising standards had a beneficial impact on their exports, which was only slightly lower than the 18 per cent that reported an adverse impact.

Trade Policy Barriers

Indian exporters face border protection regimes in foreign markets that are opaque, complex, and deceptively protectionist. Among the most pernicious forms of trade protection in horticulture are the systems of: (i) entry prices which discriminate against efficient delivery, (ii) tariff quotas that subjects imports above specified low levels to harsh tariffs, and (iii) special safeguards (SSGs) that are a source of considerable uncertainty for successful exporters. Furthermore, preferential access schemes and tariff escalation discriminate against, respectively, imports from certain sources and more advanced stages of processing. Subsidies in the form of direct payments to producers in developed countries, which have occupied centre stage in agricultural negotiations are, however, less important in horticulture. Given the other serious problems faced by Indian exporters, it is difficult to establish how far external trade regimes are today a binding constraint on India's export expansion, although there can be little doubt that these regimes would be a serious impediment if India were to emerge as a major exporter.

The three factors, logistical costs, quality, and tariffs, interact with each other and have a combined effect that is greater than the sum of the parts. Logistical costs and foreign tariffs penalize exporters multiplicatively, because the latter are imposed on a price that includes the former. Poor logistics (not just as a direct component of costs but also through waste and delay) weaken the incentives for producers to improve quality. Limited standardization makes physical inspection a precondition for transactions, creating the need for cross-hauling from producer to regional hubs to consumers, thus adding further to transport costs. Domestic protection can increase the costs of transport for exporters, because low levels of imports imply that exporters must bear the costs not only of the outward journey, but also the unutilized capacity on the inward journey.

SUMMING UP

Inefficiencies in supporting services are an important reason for the high protection that is granted to an otherwise competitive sector. The tax imposed by inefficient supporting services on the Indian horticultural producer may well offset any assistance offered by domestic protection. At the same time, inefficient logistics are also today probably as big an impediment to exports as foreign protection and restrictive standards, and could well be a bigger impediment than inefficient domestic farming. A large part of the inefficiency in services such as transport, storage and distribution is due to restrictions on domestic and foreign competition. There are also fundamental weaknesses in the infrastructure in these sectors that can be remedied by creating a favourable environment for private investment and undertaking the necessary public investment.

Looking ahead, India needs simultaneously to take steps to lower the logistical tax its farmers face at home, strengthen its regulatory framework and take an aggressive position in international trade negotiations. The inefficiencies identified in the services sectors supporting horticulture reflect problems affecting trade not only in agriculture but in all other sectors. Thus deeper reform of the entire range of services supporting agriculture, including air, maritime, and road transportation, as well as storage and marketing is vital to India's emergence as a significant trader. Enhancing the efficiency of supporting service sectors will strengthen the political case for, and economic gains from, both the liberalization of Indian agriculture trade and improved access to foreign markets.[2] At the same time, India needs to continue its active role in the WTO negotiations, seeking not just significantly lower levels of protection in destination markets, but also much greater transparency, simplicity and predictability of foreign trade regimes. While it should continue to subject foreign standards to scrutiny, and where necessary, legal challenge, there is also an urgent need to improve domestic standards and to develop mechanisms to credibly signal higher quality.

NOTES

1. The estimates of competitiveness at the commodity level are found to depend critically on the choice of reference international price. One issue is that for many horticultural products, imports are small and come from only close neighbouring countries. For example, India imports small amounts of

potatoes, mostly from its immediate neighbours (that is, Bhutan and Nepal) and imports no mangoes at all, but the most relevant external source is Pakistan. If CIF prices from these countries are used, Indian potatoes and mangoes appear only marginally competitive. But the production base of many of India's neighbours is small relative to India's needs, and any imports by India are likely to raise prices in the neighbouring countries. Thus, an NPC estimate for these commodities that takes 'neighbouring country' international prices as given is likely to be biased upward (that is, actual NPC is lower than estimated NPC). In contrast, if the prices of products from more distant and more significant sources are considered, Indian potatoes and mangoes both appear to be significantly competitive.

2. A discussion of the priorities and strategies for services reform is contained in the companion report prepared for the Indian Government, 'Sustaining India's Services Revolution'. Trade facilitation and standards in horticulture in the Indian context are being addressed by two separate reports currently under preparation.

1 An Overview

INDIAN AGRICULTURE: THE PUZZLE OF PROTECTION AND COMPARATIVE ADVANTAGE

India's agricultural policy presents a puzzle. First of all, there is persuasive evidence that India has an overall comparative advantage in agriculture (Figure 1.1a). Second, government intervention has, in the aggregate, penalized rather than assisted Indian agriculture, even though certain inputs have been subsidized. These two observations suggest that India would take a strong aggressive position in international negotiations, and would push for the elimination of trade distorting policies both abroad and at home. But the truth is quite different: India's efforts have been directed largely towards retaining the right to protect its domestic agricultural sector, and in recent years, tariffs have actually been increased on certain products. Finding the solution to this puzzle is critical to developing a case and a strategy for both domestic reforms and proactive engagement in international negotiations.

The defensive position and actions have traditionally been explained by appealing to the fact that nearly 58 per cent of India's total labour force and 75 per cent of its people below the poverty line derive their livelihood from the agricultural sector (Figure 1.1b), and liberalization could have an adverse effect on these groups. While this broad argument for caution may be valid, it lacks analytical and empirical underpinnings. After all, if the sector as a whole is internationally competitive, and can be expected to expand output in an undistorted world, then that should lead to increased employment and income for those engaged in the sector. The poor can also be expected to benefit from such an outcome, unless we have other reasons to be skeptical about the distribution of benefits from trade.

Part of the solution to the puzzle may lie in two features of Indian agriculture: *heterogeneity* of production conditions and *segmentation* of the national market. Even though India may have a comparative advantage in agriculture as a whole, certain crops, regions, and holdings are not internationally competitive. The heterogeneity has persisted because low productivity regions have been shielded from competition by policy restrictions on, and the high transport costs of, the internal movement of agricultural produce. Given the low level of diversification and the large disparity in agricultural productivity and infrastructure across Indian states, there is concern that greater openness to international trade could involve large-scale dislocation in the sector and significant adjustment costs.

Formulating a strategy that strikes the right balance between efficiency and equity will require considerable information and analysis. It would

Figure 1.1: India's Agriculture Sector

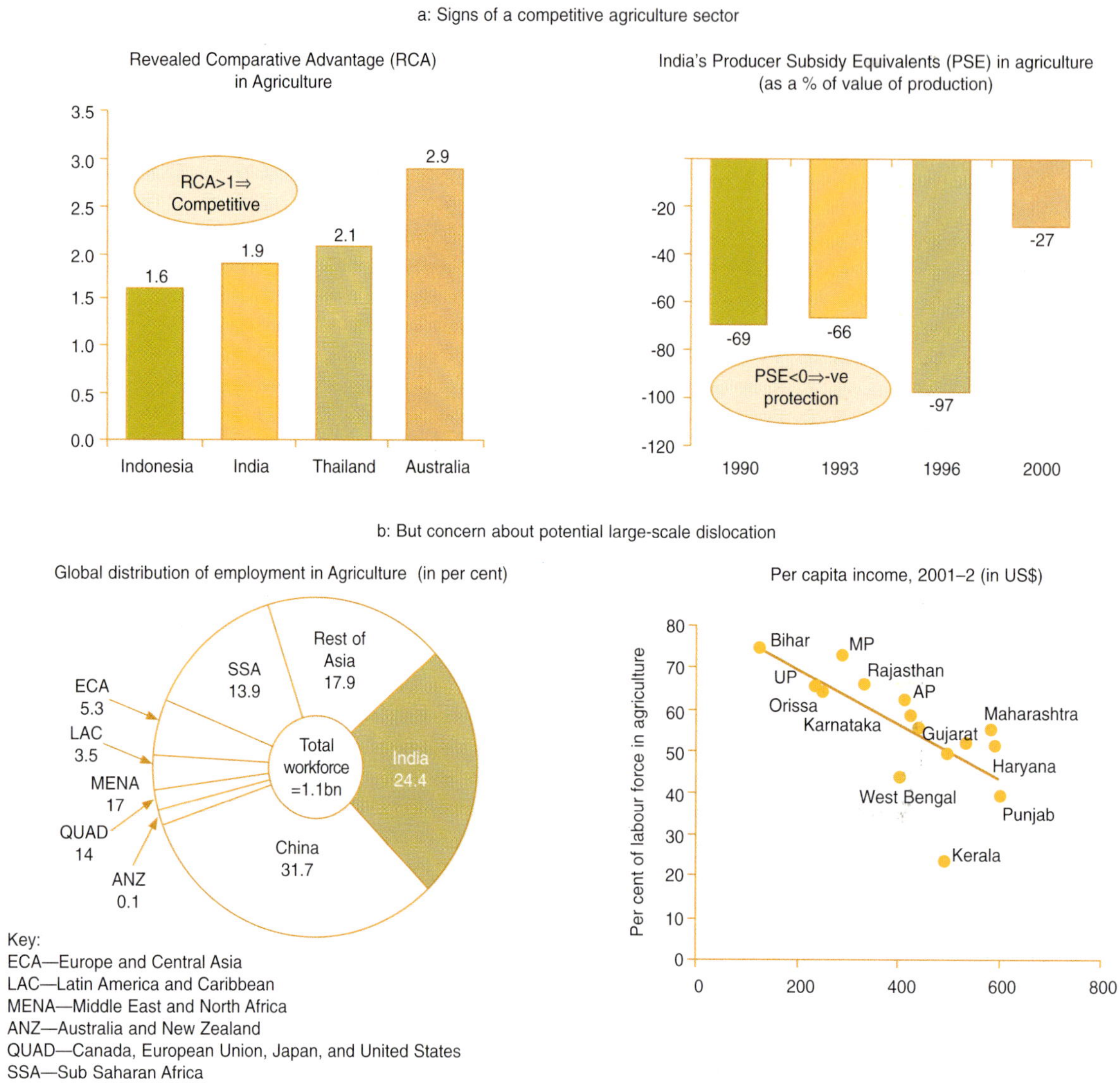

Source: Aksoy and Beghin (2004), Gulati and Narayanan (2003), GEP (2002), and Martin (2003).

involve identifying the crops and the production regions that are already internationally competitive, and the crops and regions that are not, but could be, if certain constraints were overcome through policy reforms. The strategy would also require identification of crops and regions that are inherently uncompetitive, where farmers would have to switch to alternative crops or alternative occupations in the post-liberalization period, and the associated adjustment cost involved in such a transition. Finally, welfare gains from trade liberalization would need to be evaluated under different scenarios, that is, with and without adjustment costs, as well as with and without agricultural reforms in the industrialized countries.

This study addresses some of these larger questions in the context of the Indian horticulture sector. The decision to study the horticultural sector was made (in consultation with India's Ministry of

Commerce) for three reasons. First, the sector is already among the most dynamic in both Indian agriculture and world trade. Shifting food patterns at home and abroad, based on both increased incomes and greater nutritional awareness, are steadily increasing the importance of horticultural production and trade. Second, the problems that confront perishable horticultural products, particularly relating to transportation, represent in a particularly stark form those that affect the Indian agricultural sector as a whole. Finally, the horticulture sector may represent a 'picture of the future' for agriculture as a whole because it is a sector which is less restricted by measures such as subsidies and (arguably) tariffs, but sees an increasing demand for higher standards due to health, safety, and broader quality concerns. Thus the sector encompasses the major challenges that Indian agriculture faces in a world of increasing incomes, tougher delivery schedules, less restrictive conventional trade barriers, and more demanding product standards.

HORTICULTURE: POTENTIAL AND PROBLEMS

The performance of the Indian horticultural sector in the last decade has provoked dramatically differing assessments.[1] Some have described it as a 'golden revolution', presumably to distinguish it from the earlier 'green revolution' (Bannerjee 2005). Others have dismissed these views as a hyperbolic description of a fundamentally stagnant sector. We find that the truth from a trade perspective is more complex. The sector is beginning to come alive but it faces formidable constraints.

India is one of the largest and lowest cost producers of high value agricultural commodities and yet has a minuscule share in global trade. It produces nearly 11 per cent of all vegetables and 15 per cent of all fruits in the world (Figure 1.2). The unit values of its exports (free on board, FOB) are nearly half the corresponding world unit values. Yet its share in global exports of vegetables is only 1.7 per cent and in fruits a meagre 0.5 per cent. At the same time, its own market is heavily protected. Finding an explanation for this strange conjunction of low costs, low exports, and high protection is vital to any assessment of the competitiveness of Indian agriculture in a liberalized global trading environment.

Many economists and policy makers have, in the past, attributed India's poor export performance to an overvalued real exchange rate and the anti-export bias in its trade policy, the latter being a by-product of the Government of India's (GoI's) broader objective of achieving self-sufficiency and food security in agriculture. The gradual depreciation of the Indian currency since the mid-1980s, the large devaluations in 1991 and 1993, and the subsequent move to a flexible exchange rate regime corrected the overvaluation in the real exchange rate. During the last several years, a number of policies have been initiated to reduce the anti-export bias in India's trade policy. While these reforms have relaxed some of the constraints affecting the horticulture sector, the much anticipated surge in export is yet to materialize. While India's unit value of exports continues to be one of the lowest in the world, its share in global exports in recent years has been less than that in the early 1990s (Figure 1.2).

A complete understanding of India's competitiveness in high value agricultural commodities has proved elusive because the sector has not been subject to an integrated analysis—from farm to retail. Most studies have examined only a single stage of the supply chain, for example, productivity at the farm, efficiency of agricultural markets, or problems associated with poor infrastructure, in isolation of

Figure 1.2: India's Cost of Production and Market Share

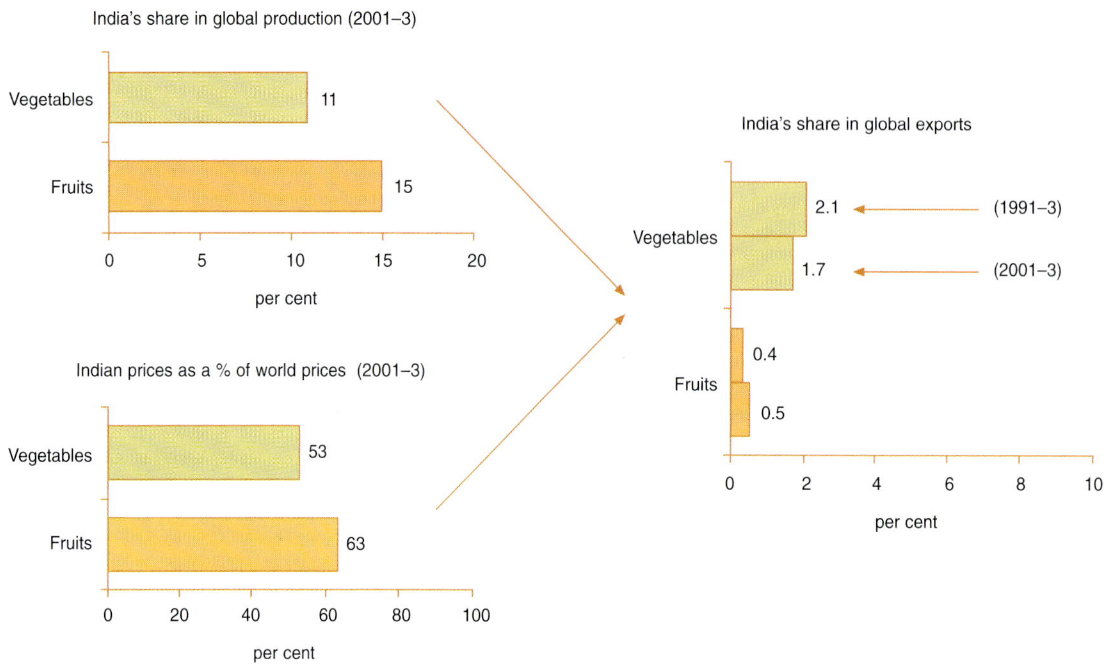

Source: UN COMTRADE Statistics, Staff estimates.

the rest of the supply chain. The researchers were not in a position to assess how the problems they identified at the single stage of the supply chain compared with the (unidentified) problems in the rest of the chain.

This study undertakes a supply chain analysis of thirteen high value agricultural commodities, based on primary surveys that covered 1400 farmers, 200 commission agents, and 65 exporters across fifteen different Indian states (see Appendix 2 for details). It adds to the existing literature by laying out the cost and price structure of all agents and all markets in the supply chain, thus providing an integrated view of the sector. It becomes possible to analyse the relative importance of the factors that put Indian produce at a disadvantage relative to that from other countries. This framework also enables us to simulate the potential costs and benefits of policy intervention at different stages of the supply chain and thereby help prioritize and formulate policies that would have the biggest impact on India's competitiveness.

WHAT IMPEDES EXPORTS?

The supply chain analysis indicates that from a purely cost perspective the biggest impediments to the competitiveness of India's horticultural exports lie outside the sector rather than inside it (Figure 1.3).[2]

Factors within the agriculture sector are found to have a limited bearing on competitiveness from a trade perspective, with the impact on final price seen to be much smaller than the factors outside the sector. For example, the average price at the farm-gate for a typical horticulture product is found to be only 12–15 per cent of the price at which the

Figure 1.3: Barriers to Trade: External vs Domestic Factors

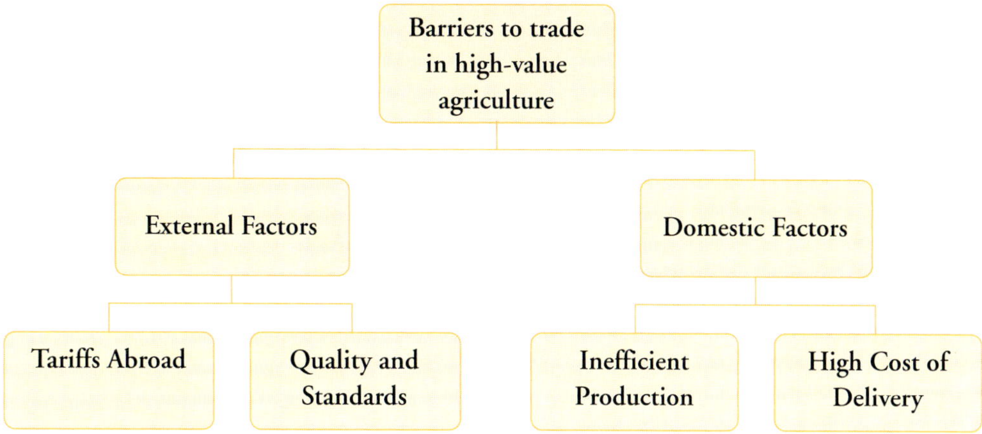

Source: Bank Staff.

product is retailed in the destination market. This implies that a 20 per cent improvement in yields (which is huge, since yields have increased by only eight percentage points during the last 15 years!) will lead only to a 2.4–3 percentage points reduction in the final price. This is not to understate the importance for Indian farmers of developing better varieties of horticulture products that are internationally more competitive on the quality dimension. Improved agronomics and better technology that can enhance the quality of products in terms of their appearance, size, colour, taste, and shelf-life are of considerable significance if India's horticulture sector is to become competitive both at home and abroad. But since the production-related problems in agriculture are well-known and extensively researched,[3] they do not form the major focus of this study.

India's exports face three major impediments. The first is the 'logistical tax' imposed by high delivery costs which significantly erode the production cost advantage enjoyed by Indian farmers. The second is the gap between the high standards required by governments and buyers, especially in richer countries, and the low standards and weak conformity assessment mechanisms in India. The third problem is the trade policy barriers that Indian exporters face in foreign markets.

Comparing India to selected countries, across destination markets, modes of transport, and commodities traded indicates that, on average, India's international transportation costs are 20–30 per cent higher than those faced by other countries. Since transportation costs are found to account for nearly 45 per cent of retail price when air transport is used and 25 per cent in case of maritime transport, Indian products are 5 to 15 per cent more expensive than their foreign counterparts simply on account of high international transportation costs. Logistical cost is affected by two factors: geographical distance which cannot be remedied and logistical inefficiency which can but has not been remedied. Our estimates suggest that an improvement in the efficiency of the transportation and logistics system (both the domestic as well as the international transport costs) by 20 per cent could reduce final price by as much as twelve percentage points (that is, a saving of Rs 8 on a final price of Rs 62).

Poor quality has two dimensions: weak safety and health standards that legitimize prohibitive SPS barriers imposed by foreign governments; inadequate standardization of products that reduces attractiveness for foreign retailers. Our survey of exporters suggests that Indian products are not perceived to be of lower quality by Indian exporters. Furthermore, many Indian exporters have focused on regional markets and so far there have been limited encounters with the demanding standards of the industrial countries—described as a pattern of 'defensive commercialization' in a recent World Bank study on the impact of standards on horticultural trade. There is clear evidence that stringent industrial country standards have been a prohibitive barrier for some products such as mangoes. In the case of other products such as grapes, exporters have incurred temporary losses due to the rejection of consignments that do not meet standards. In general, producers who have had to make adjustments in response to higher standards have typically incurred significant consignment-specific or recurrent transactions costs. Meeting increasingly high standards will be a critical challenge in expanding exports to more profitable destinations.

Indian exporters face border protection regimes in foreign markets that are opaque, complex, and deceptively protectionist. Among the most pernicious forms of trade protection in horticulture are the systems of: (i) entry prices which discriminate against efficient delivery; (ii) tariff quotas that subjects imports above specified low levels to harsh tariffs; and (iii) SSGs, which are a source of considerable uncertainty for successful exporters. Furthermore, preferential access schemes and tariff escalation discriminate against, respectively, imports from certain sources and more advanced stages of processing. Subsidies in the form of direct payments to producers in developed countries, which have occupied centre stage in agricultural negotiations are, however, less important in horticulture. Given the other serious problems faced by Indian exporters, it is however difficult to establish how far external trade regimes are today a binding constraint on India's export expansion although there can be little doubt that these regimes would be a serious impediment if India were to emerge as a major exporter.

The three factors, tariffs, logistical costs, and quality, interact with each other and have a combined effect that is greater than the sum of the parts. Logistical costs and foreign tariffs penalize exporters multiplicatively, because the latter are imposed on a price that includes the former. Poor logistics (not just as a direct component of costs but through waste and delay) weaken the incentives for producers to improve quality. Limited standardization makes physical inspection a precondition for transactions, creating the need for cross-hauling from producer to regional hubs to consumers, thus adding further to transport costs. Domestic protection can increase the costs of transport for exporters, because low levels of imports imply that exporters must bear the cost not only of the outward journey, but also the unutilized capacity on the inward journey.

WHY ARE DOMESTIC TARIFFS SO HIGH?

In order to assess the competitiveness of Indian horticulture, we estimated the NPC, which is the ratio of domestic to international prices (appropriately modified to take into account transport and other costs), at the all-India level, both separately and aggregated over all regions and over all months, for the period 1997–2002.

Addressing this question reveals the inadequacy of analyses based solely on aggregative data. For example, data limitations preclude effective control for differences in quality and for fine distinctions between varieties between foreign and domestic commodities. Nevertheless, our results support the findings of a number of previous studies that India's horticulture sector is competitive in terms of prices. As Figure 1.4 shows, the average NPC has been consistently below one between 1997 and 2002; there is some variation across products (discussed in Chapter 3). In the case of bananas, domestic prices on average are consistently less than half the price of imports when they arrive in local markets. Grapes and oranges have slightly higher NPCs, but both appear to be becoming more competitive over time. Apples and mangoes, however, appear to be borderline competitive. For mangoes, the NPC was significantly higher than one only in 1998, though it has been fluctuating close to that level. All vegetables are also competitive except for potatoes in some years.

Why then does India have such high tariffs for even the horticulture products that are clearly competitive? There could be several explanations for this. One is that the tariffs are purely precautionary, and meant to deal with circumstances where foreign suppliers benefit from a sudden reduction in costs, thanks to an act of God or government. However, we find that international unit prices of horticultural products have not displayed much variability, and governments have been remarkably restrained in subsidizing such produce. Another explanation is that the market for a product is not homogeneous but differentiated, either horizontally (that is, foreign and domestic products are different varieties) or vertically (that is, foreign and domestic products are of different quality), in which case the tariff can serve a protectionist purpose. While this explanation cannot be ruled out, our survey of exporters does not suggest that Indian products are significantly inferior to foreign products.

We attach greatest credence to a combination of two explanations: tariffs are intended to protect domestic produce in all regions and all seasons regardless of the variations in prices across space and time; and that the level of tariffs is not always

Figure 1.4: NPCs for Horticulture: All India, 1997–2002

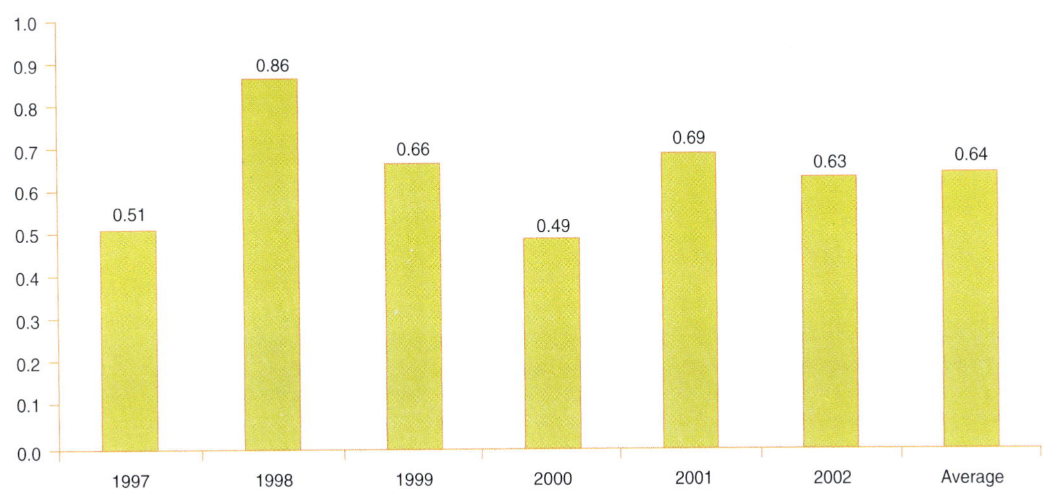

Source: Staff estimates (see Appendix for a discussion of the estimation methodology).

sensitive to the specific current circumstances of individual products, but is set at roughly similar, sometimes historically determined levels for a wide range of products—a form of policy hysteresis.

A better understanding of the aggregate numbers can be obtained by looking at the evidence across regions and over time (Figure 1.5). First of all, we find large regional variations in prices. This is a consequence of the vastness of India, where distances between growing and consuming regions are often quite large. These large distances become even more of an issue because of high domestic transportation costs, and high wastage when commodities are

Figure 1.5: Price Variation across Space and Seasons for Domestic and Imported Apples and Potatoes

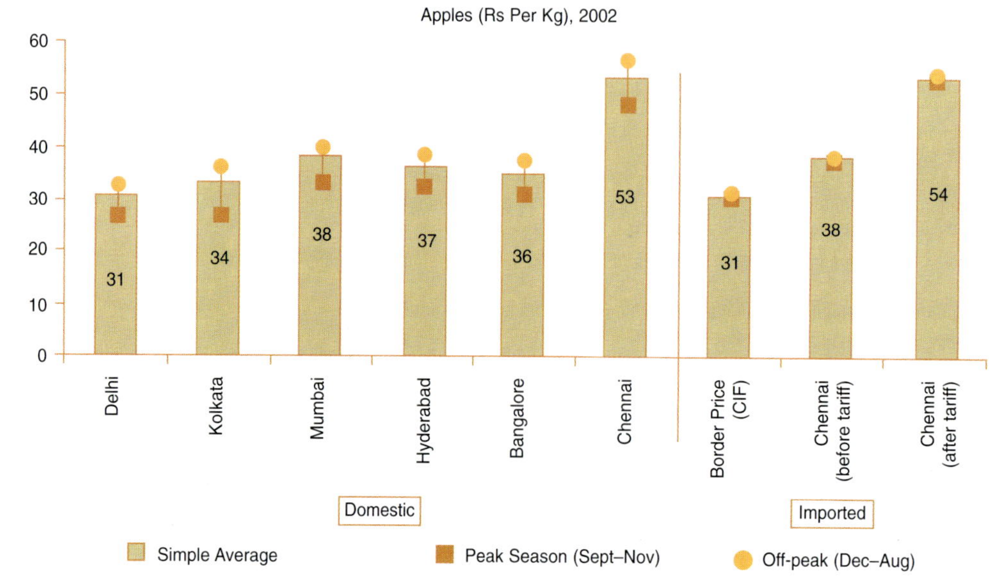

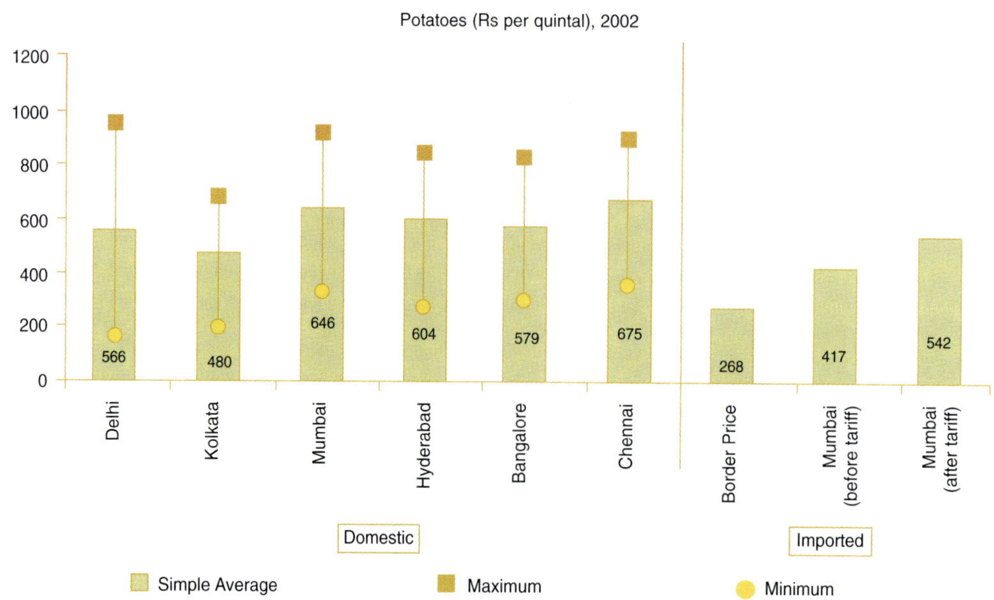

Source: NHB Survey, Staff estimates.

moved, due in large part to poor infrastructure and the lack of cold storage infrastructure. Consider the case of apples. The major apple producing areas of the country are located in the north and north-eastern parts of India, and most apple trade happens through Delhi. Domestically grown apples are much cheaper than foreign apples in markets located near these regions. However the high cost of transporting and marketing these in the southern and western regions leads to very high prices in these markets, rendering domestic production uncompetitive. Apple prices in Chennai are almost 70 per cent higher than those in Delhi. Similarly, the largest potato growing states are Uttar Pradesh and West Bengal, together accounting for more than 70 per cent of production. There is a 40 per cent price differential between Kolkata in the east and Chennai in the south. This margin makes imported potatoes relatively more expensive in the north and east, but relatively cheaper in the western and southern regions.

There are also large variations in prices and availability across time because production of many horticultural commodities is seasonal in nature. In India, seasonality effects are further accentuated as storage facilities are both expensive and rare. Still, many of the products considered in our study were found to be competitive, were available throughout the year domestically and not imported at all, and would probably not be imported in large amounts even if tariffs on them were brought down to zero. These include products such as bananas, onions, and tomatoes. But in the case of some products, there were seasons in which prices soared above international levels or availability declined drastically. In Delhi, for example, potato prices ranged from Rs 2.91 per kg to Rs 9.53 per kg in a single year (2002). Regions also differ widely from each other in terms of availability. Apples are available almost throughout the year in Delhi, but are missing from Chennai markets for four months of the year. Grapes are available for eight months of the year in Delhi, but for only five months in Kolkata. In general, availability of fruits and vegetables tends to be better in the north, and worse in the south.

It would clearly be efficient to allow consumers in all regions and at all times to buy from the cheapest domestic or foreign source. Current high tariffs, however, penalize consumers in regions located far from production sources and in seasons far from the harvest season. While the net costs of protection are well understood, it is notable that tariffs restrict imports in places and during times which domestic products find difficult to reach. The existence of tariffs in such circumstances creates an incentive for inefficient transport and inefficient storage. The result is that tariffs hurt consumers, and they subsidize an inefficient distribution and storage system more than they subsidize producers. Therefore, the net welfare loss they create is even larger than usual.

To conclude this overview, the high costs of product delivery are today probably as big an impediment to exports as foreign protection and demanding standards, and could well be a bigger impediment than inefficient domestic farming. They also increase the cost of supplying the domestic market and are probably an important reason for the high protection that is granted to an otherwise competitive sector. A large part of the high costs is attributable to weaknesses in the transport, storage, and marketing infrastructure that only significant public and private investment can remedy. However, costs are also kept unnecessarily high by policy restrictions, particularly those that impede competitive supply of transport and distribution services. Remedying the acute inefficiencies of supporting service sectors will greatly strengthen both the political case and economic gains from further liberalization of Indian agriculture.

NOTES

1. For the purpose of this study, the horticultural sector is defined to include fruits and vegetables, which consist of the following four main categories: (i) fresh fruits; (ii) fresh vegetables; (iii) processed fruits; and (iv) processed vegetables. Fresh vegetables and fresh fruits include products that are only found in Chapters 7 and 8 respectively of the HS Nomenclature while 'processed' fruits and vegetables refer to products found only in Chapters 20 and 21.
2. The barriers to trade mentioned in Figure 1.3 are only a subset of factors affecting India's trade in high-value agriculture. For a more detailed discussion of potential barrier to trade, see World Bank (2005a).
3. See Gulati and Pursell (2003), Gulati, Pursell, and Mullen (2003), Kapila and Kapila (2002), and Umali-Deininger (2004).

2 International Trade in Horticulture[1]

INDIA'S HORTICULTURE FROM A GLOBAL PERSPECTIVE

The horticulture sector in India is an important segment of the economy. In 2003–4, it contributed nearly 30 per cent of agricultural gross domestic product (GDP) from only 8.5 per cent of India's gross cropped area. Its growth in India has been steady but not spectacular. World production of fruits and vegetables grew by 30 per cent between 1980 and 1990 and by 56 per cent between 1990 and 2003, reaching 1274 million tonnes by 2003. As Figure 2.1 shows, much of this growth occurred in China, where production grew by 134 and 236 per cent during the past two decades, respectively. China is currently the world's largest fruit and vegetable producer, with a share of 34 per cent, followed by the Latin America and Caribbean region (11 per cent), India (10 per cent), and Africa and the EU (both at 9 per cent).[2]

A comparison of productivity level and growth across selected countries shows that India suffers from low productivity, and improvements in yields have been considerably lower relative to other countries, especially China (Figure 2.2). In fruits, all countries, Brazil, Chile, India, China, and South Africa, have seen improvements in yield per

Figure 2.1: Production and Growth of Horticulture in Selected Countries, 1980–2004

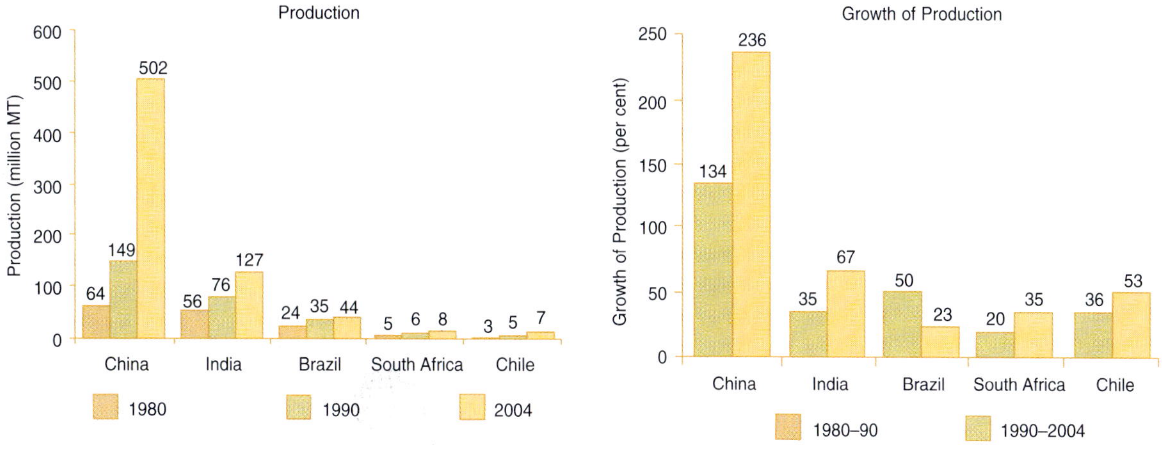

Source: FAO, Staff estimates.

Figure 2.2: Yield per Hectare in Fruits and Vegetables in Selected Countries, 1990–2004

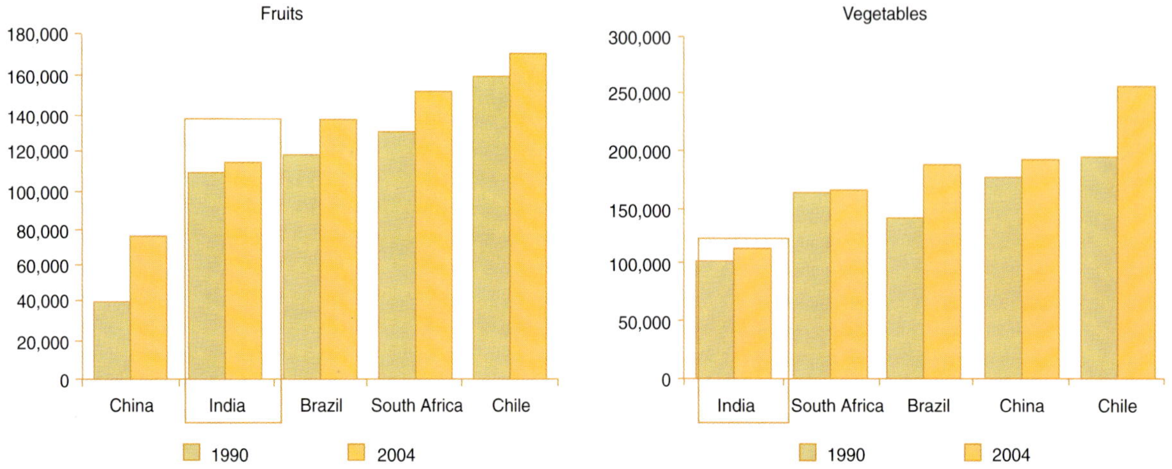

Source: FAO, Staff estimates.

hectare, though India's improvement has been the lowest in the group. While Chile and South Africa are the clear leaders in terms of absolute yield per hectare, China has seen the biggest improvement in yield between 1990 and 2004. Despite a smaller improvement, India has maintained its fourth place. In vegetables, the picture is similar to that in fruits, except that the yield gap between India and the other countries is considerable and is widening rapidly (except with South Africa). For example, in 2004, while the yield level in India was nearly 70 per cent of the Chilean level in case of fruits, it was only 45 per cent in case of vegetables. These figures suggest that with adequate investment and the right policies, India has the potential to raise its yield in horticulture.

The trend growth of overall Indian horticultural production has been 4.2 per cent per annum over the period 1991–2 to 2001–2 (Figure 2.3). Vegetable production has grown a little faster (4.6 per cent per annum) than fruit production (3.6 per cent per annum). Growth in vegetable production has come from both an increase in acreage (2.1 per cent per annum) and an increase in land productivity (2.5 per cent per annum). However, the growth in fruit production is attributable much more to an expansion in acreage (3 per cent per annum) than a spurt in productivity (0.6 per cent per annum). In each case, the most recent year or two has seen a slight dip in both production and productivity after a decade of steady growth.

INTERNATIONAL TRADE

Trade in horticulture products has been among the most dynamic areas of international agricultural trade. It has been stimulated by rising incomes and growing consumer interest in product variety, freshness, convenience, and year-round availability. A facilitating role has been played by advances in production, post-harvest processing, and logistical technologies, and increased levels of international investment (Diop and Jaffee 2004). As indicated earlier, India however has not emerged as a major player in international trade, and its exports are a small fraction of both world horticultural exports and domestic production.

Figure 2.3: Decomposition of Growth Rate of Horticulture Sector
(Growth due to increase in area and growth due to increase in productivity)

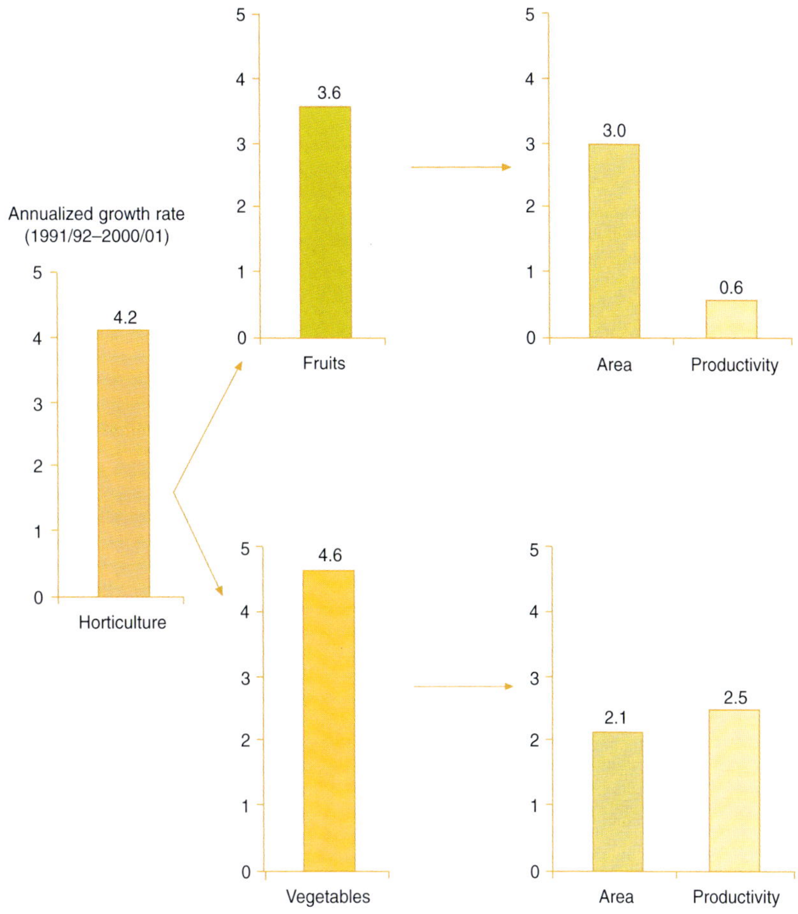

Source: FAO, Staff estimates.

Global Trade Patterns

The structure of world trade in fruits and vegetables does not reflect the structure of production. Many of the largest producers are not significant traders due to a combination of domestic demand and geographical/logistical factors. For instance, only a small percentage of China's and India's production is exported due to a strong domestic demand, fuelled by growing income and a large and rapidly growing urban population. For example, the share of fruits and vegetables in the Indian rural consumption basket rose from 7 per cent in 1972–3 to 13 per cent in 1999–2000, and from 10 to 16 per cent in the urban consumption basket over the same period. In contrast, Latin American countries (for example, Mexico, Chile, and Costa Rica) are among the world's leading exporters of fruits and vegetables mainly due to their proximity to the large US market.

European Union is the world's largest market and supplier of fresh and processed fruits and vegetables, accounting for 51 per cent of world imports, and 39.5 per cent of world exports. EU trade in fruits and vegetables is, however, largely intra-regional.

Intra-EU imports represent 64 per cent of EU imports, while 83 per cent of EU exports occur between its 15 member states. Still, with its affluent and ageing population and its high factor costs and cold winter, this region represents one of the largest fruit and vegetable markets for non-EU countries ($13.2 billion).

Intra-regional trade is also significant in the world's second largest fruit and vegetables market—North American Free Trade Agreement (NAFTA). Trade between Mexico, Canada, and the US accounted for 49 per cent of NAFTA's imports and 53 per cent of NAFTA's exports in 2001. Intra-NAFTA trade is most important for fresh vegetables. For this commodity group, 90 per cent of exports and 86 per cent of imports occur within the trade group (Huang 2004).

Interestingly, thanks to growing incomes in the 1990s, middle-income countries have seen their own markets become major destination of exports from other countries, with import demand totalling $11 billion in 2001. South–South trade—that is, trade between developing countries, excluding China and India—totalled US$5.4 billion in 2001, accounting for 45 per cent of developing countries' imports. Japan has also emerged as a significant market for fruits and vegetables over the 1990s, with import demand reaching $5.8 billion in 2001.

While many countries have entered the fresh fruit and vegetable export markets, only a few middle-income countries have succeeded on a sustained basis. The average shares of developing countries in world exports of fresh fruits and vegetables hide the heavy domination of trade by just a handful of middle-income countries. Between 1997 and 2001, just four countries, Mexico, Chile, Ecuador, and Costa Rica, accounted for 43 per cent of world developing country exports of fresh fruit (FAO 2003) and the top four suppliers, Mexico, China, Argentina, and Syria accounted for 67 per cent of fresh vegetable exports by developing countries (FAO 2003).[3] A very small number of medium-income countries have been successful in the processed segment of the export market, and developing countries as a group account for a relatively low share in the exports of processed products (36 per cent in 2001). China, Thailand, Chile, and Turkey account for 58 per cent of developing countries' exports of processed fruits and vegetables (FAO 2003).

Role of Geography

The powerful influence of geography on trade in horticultural products is evident in global trade patterns. Figures 2.4a and 2.4b show, for example, that a large proportion of the imports of US comes from proximate countries and a large portion of India's exports goes to neighbouring countries.

Geography is evidently a barrier to India's trade in horticulture. India is located in one of the poorest regions of the world, and is disadvantaged relative to some of its competitors in terms of distance to the more attractive export markets like the US and EU. As shown in Figure 2.4b, distance of a region from India is a strong predictor of its share in India's exports.[4] Holding everything else constant, India's share of exports to any destination market is found to decline by ten percentage points for every 1000 kilometres increase in the distance to a market, and any market that is beyond 14,000 kilometres from Indian borders is unlikely to be served by Indian exporters.

However, distance is important but not decisive. A number of countries from the Cairns group of

Figure 2.4: Distance has a Considerable Influence on US Imports and India's Exports, but Chile's Products seem to have Broken the Distance Barrier

a: US's horticulture imports are influenced by distance .....

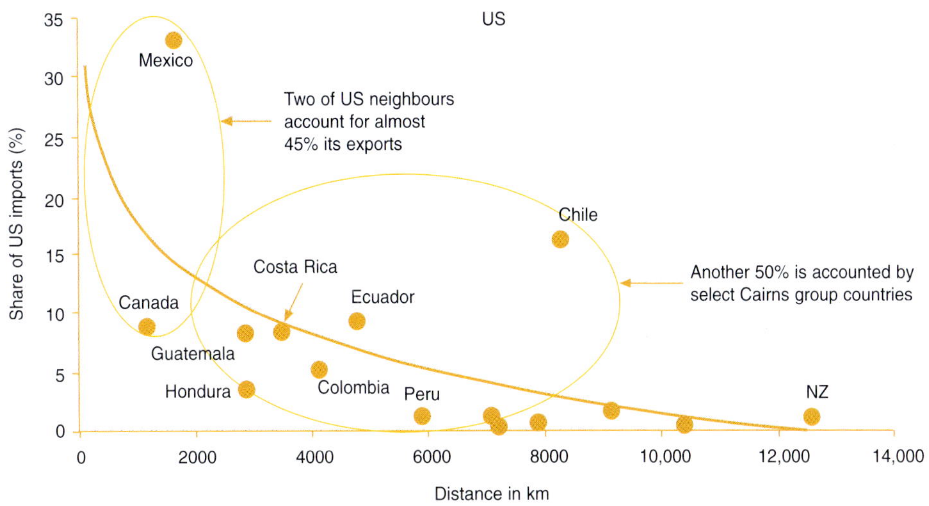

b: ... And India's horticulture exports are significantly influenced by distance .....

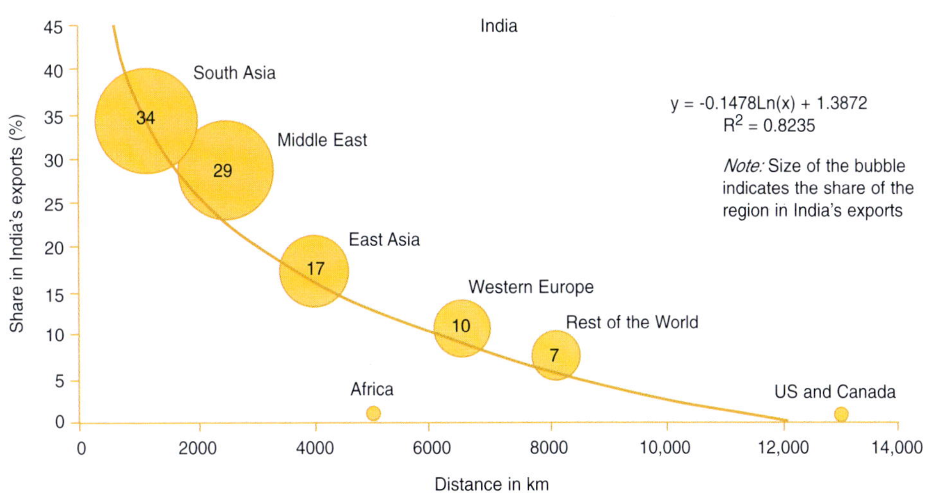

agricultural exporters have succeeded in establishing a significant presence in relatively remote markets and Chile stands out as the most successful example of a developing country that has 'broken the distance barrier.' As shown in Figure 2.4c, 80 per cent of Chile's horticulture exports are transported to destination markets that are 9000 kilometres or more away from its borders. The corresponding figure for India is 1 per cent. So while India's export performance is considerably influenced by distance, Chile's is not.

In Chapter 4, we argue that the improvement of international and domestic logistics can cause 'effective distance' to shrink and diminish, dampening

c: ... but Chile's exports appear almost invariant to distance

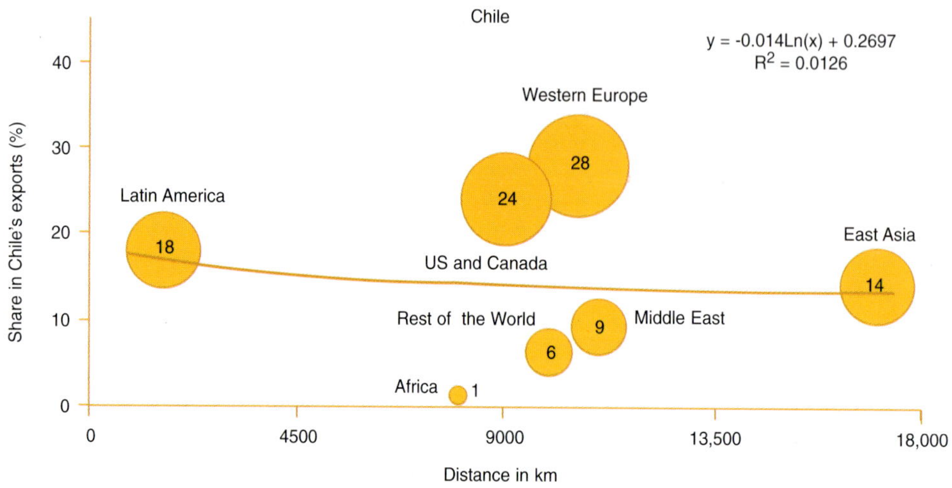

Source: UN COMTRADE Statistics, CEPII (www.cepii.fr/anglaisgraph/bdd/distances.htm), Staff estimates.

the influence of remoteness on exports. Chile's adoption of an open seas and open skies policy, and Kenyan reform of air transport have certainly played a role in their emergence as horticultural exporters.

Quality of Indian Exports

Quality is an important determinant of trade in horticulture and better quality products (in terms of appearance, taste, size, and shelf-life) tend to attract considerable premium in the global market. The perception of Indian exporters about the quality of Indian exports relative to foreign products indicates that Indian products could be potential candidates for the upper end of the quality spectrum. On a scale from 0, signifying 'far inferior to comparator', and 10, signifying 'far superior to comparator', each Indian product scored between 6.7 and 7.9 when compared with products in the destination market. While there may be an upward bias in these assessments, they do seem to refute any notion of quality pessimism about Indian horticultural products.[5] At the same time, the fruits and vegetables being exported from India are perceived to be better than those sold domestically in terms of all the quality dimensions, indicating that higher quality products are being exported (Figure 2.5).

A Closer Look at India's Trade

India is not a major player in international trade, and its exports are a small fraction of both world horticultural exports and domestic production. India's horticultural exports have hovered around 20 per cent of its total agricultural exports (Figure 2.6). However, onions are the only horticultural product where exports are a significant proportion of domestic production, with the percentage varying between 5 and 12 per cent over the period 1990–1 and 2001–2. The only other product where exports are more than 1 per cent of aggregate production is grapes, with the percentage varying between 1 and 3 per cent over the same period. Imports of horticultural products have declined from 50 per cent of total agricultural imports during the early 1990s to 25 per cent in 2003. Imports are a tiny fraction of domestic production in all cases, including peas. Bananas and oranges among fruits, and peas among vegetables, are the only products whose exports are closely correlated with production.

Figure 2.5: The Perception of Exporters about the Quality of their Exports

Source: Exporters' Survey, Staff estimates.

Figure 2.6: Exports and Imports in Agriculture and Horticulture

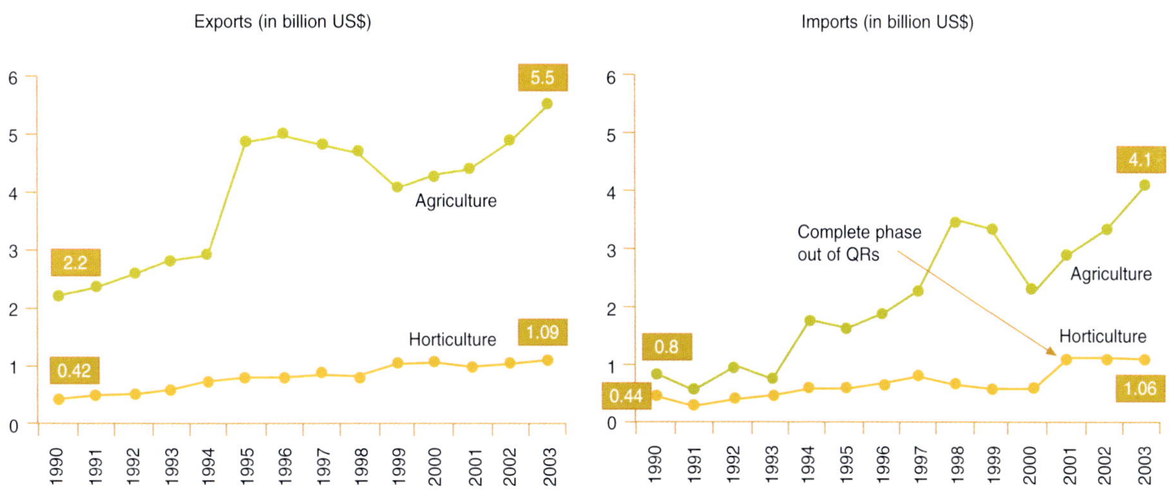

Source: UN COMTRADE Statistics, Staff estimates.

On the whole, India is a 'net exporter' of all major horticulture products.

The most important items among India's horticultural exports are fresh onions, mango pulp and fresh mangoes, dried walnuts, and fresh grapes. Dried and preserved vegetables account for a quarter of all horticultural exports. Far more striking is the regional composition of India's exports. As indicated in Figure 2.7, a significant proportion of India's exports of fresh fruits and vegetables goes to countries in South Asia and Middle East. Only a

Figure 2.7: Destination of India's Exports, 2003

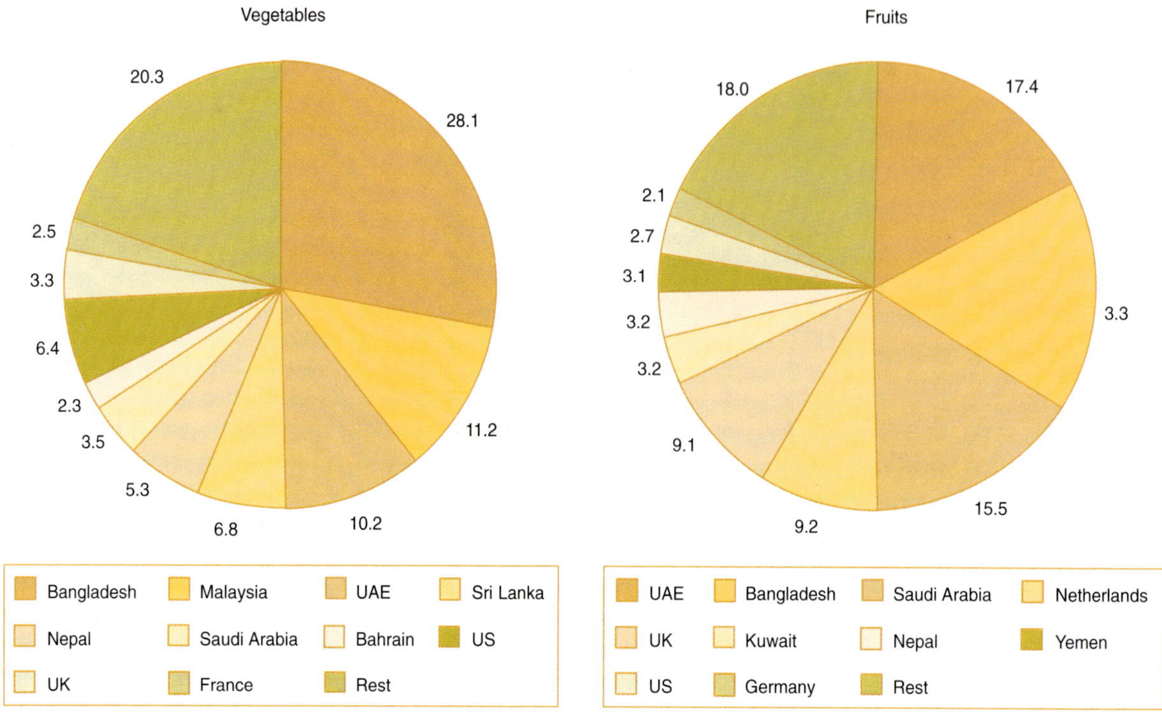

Note: Fruits excluding nuts.
All figures in percentage.
Source: UN COMTRADE Statistics, Staff estimates.

small proportion of India's exports goes to industrial countries, particularly those that are home to a large Indian diaspora. In contrast, a large part of India's limited exports of processed fruits and vegetables goes to industrial country markets.

Exports of vegetables and fruits have grown in the year 2002–3. The fastest rates of growth have been recorded by tomatoes, peas, dried vegetables, guavas, mangoes, onions, and apples. But India's share of world exports is still small—the share of vegetable exports peaked at around 2 per cent in 1992, declined to less than 1.5 per cent in the mid-1990s, and recently recovered to around 1 per cent. Its share of world exports of fruits is even smaller, less than 1.25 per cent of world trade in recent years. There are relatively few commodities—green peas and mangoes—where India is a relatively large exporter, in the sense of facing a relatively low own-price elasticity of demand (FAO 2004).

Between 1990 and 2003, the dollar value of imports of all horticultural products more than tripled in value, growing from an average of about $440 million to $1.06 billion (and from $59 million to $178 million for the commodities under study). Most of this increase came after the phase-out of quantitative restrictions (QRs) starting in the second half of the 1990s. Imports of fresh fruits, especially apples, have risen considerably, growing by more than 245 per cent per annum between 1997 and 2003, albeit on a small base. Growth has also been rapid for certain processed food products such as apple juice and dried vegetables, though the single

Figure 2.8: Exports and Imports in Agriculture and Horticulture

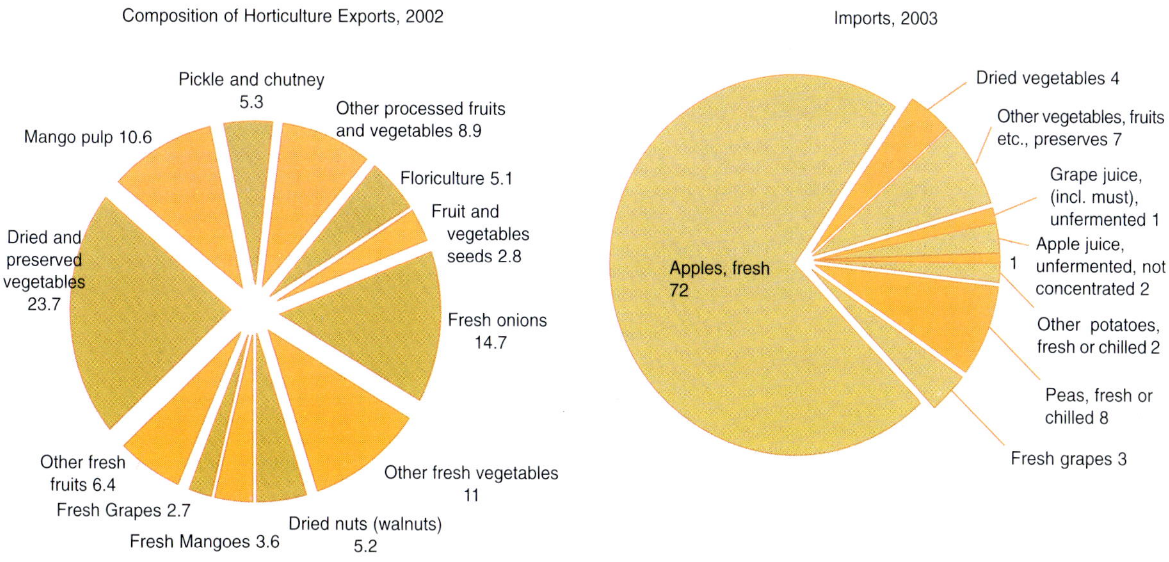

Notes: All figures are percentages.
This figure is based on import data for a subset of commodities in the horticulture sector.
Source: UN COMTRADE.

most dominant item in this category continues to be dried peas (Figure 2.8). In 2003, imports of dried peas amounted to $162 million. Imports of fresh vegetables are small, mostly of green peas. However despite the recent increase, import penetration remains insignificant. For fresh fruits, it is less than 0.1 per cent of domestic production, and is also negligible for most fresh vegetables.

Import volumes have risen at a time when world prices are stable or rising. For example, in the case of dried peas, periods of high imports have been associated with higher prices. Similarly, for both apples and grapes, the fastest growing items in the fresh fruits category, unit prices have been higher on the average in the post-2000 period, than in the decade preceding it. From this it seems clear that increasing import volumes are more reflective of demand conditions in India, than changes in international supply.[6]

NOTES

1. This Chapter draws heavily upon Diop and Jaffee (2004), FAO (2004), Regmi and Gehlhar (2005), and World Bank (2004, 2005a).
2. According to India's National Horticulture Board, India is the second largest producer of fruits and the largest producer of vegetables in the world.
3. The bulk of Syria's trade is with other Middle Eastern countries while that of Argentina's is targeted primarily towards other Latin American countries.
4. The same does not hold true for manufacturing exports. If we replace the share of India's exports in horticulture by manufacturing goods in the y-axis of Figure 2.4b, the trend line becomes an upward sloping curve indicating that distance is not a major determinant of the destination of exports in case of manufacturing goods.
5. However, it is important to note that a large part of Indian horticulture exports either go to nearby countries in Asia or the Middle East or to ethnic

stores in Europe and North America. It is well documented that the retail customers in ethnic stores are mostly Indian diaspora who give preference to the type of vegetable and its origin over the quality and accept certain variations in vegetable colour, texture, blemishes, and freshness that other 'mainstream supermarkets' retail shoppers might reject (World Bank 2005a).

6. Some economists have argued that India's agricultural surpluses and deficits are potentially large relative to the level of world trade and thus international prices are likely to rise when India enters the market as a buyer and fall when India enters as a seller. Chand (2001) uses India's trade in wheat to illustrate this hypothesis.

3 Is there a Rationale for Protection?

INTRODUCTION

Indian policymakers have been cautious in opening up the horticultural sector. Until the mid-1990s, India was almost closed to trade. Imports of most agricultural products were controlled through a licensing regime as well as quotas. Exports were canalized through designated entities, and important export items like onions were subject to quantity controls. Starting in the late 1990s, India started easing many of its export controls. It also started to gradually phase out quantitative controls on imports from 1996 as part of its WTO commitments. But while quotas were being withdrawn, tariffs on some items were raised, and most tariffs now range between 30–50 per cent for horticultural products (Table 3.1).

Table 3.1: India's Applied Tariffs for Main Crops, 1990–2002

(simple average in %)

HS-6	Product	1990	1992	1997	1999	2000	2001	2002
070190	Other potatoes, fresh or chilled	..	..	10	15	38.5	35	30
070200	Tomatoes, fresh or chilled	..	..	..	..	38.5	35	30
070310	Onions and shallots, fresh or chilled	..	..	10	..	38.5	35	30
070810	Peas, fresh or chilled	10	10	10	15	38.5	35	30
071290	Dried vegetables	..	..	10	15	35	35	30
080300	Bananas, including plantains, fresh	..	..	..	40	38.5	35	30
080450	Guavas, mangoes, and mangosteens, fresh	..	..	40	40	38.5	35	30
080530	Lemons and limes, fresh or dried	..	..	..	40	38.5	40	40
080610	Fresh grapes	30	..	30	25	38.5	35	40
080810	Apples, fresh	..	..	..	40	50	50	50
200190	Other vegetables, fruits, etc., preserves	..	..	40	40	38.5	35	30
200950	Tomato juice, unfermented, not contaminated	..	..	..	40	38.5	35	30
200960	Grape juice, (incl. must), unfermented	..	..	40	40	38.5	35	30
200970	Apple juice, unfermented, not contaminated	..	..	40	40	38.5	35	30

Sources: UNCTAD TRAINS database, WTO IDB CD ROM 2004.

In this section, we examine the price competitiveness of a select basket of horticultural commodities to assess India's competitiveness in these products, and the rationale behind India's tariff policy. Since the aim is to look at the potential disruptive effects of opening up on the sector as a whole, the choice of commodities has been made so as to cover the most widely grown crops among fruits and vegetables (see Appendix 2). The chosen basket of products includes apples, bananas, grapes, mangos, and oranges in fruits, and tomatoes, potatoes, onions, peas, and okra among vegetables. In fruits, this selection covers about 66 per cent of total production by volume. For vegetables this figure is 54 per cent. It must be borne in mind, however, that these selected commodities are not necessarily the products where India has the greatest export advantage, or those where it faces the greatest import competition.

The price competitiveness of a commodity is measured by estimating the Nominal Protection Coefficient (NPC). This is defined as the ratio of the price of domestic produce to the price of imported products, after taking account of transportation and other marketing costs. An NPC greater than one thus implies that domestic prices are higher than international prices. This measure has been employed in previous studies (Gulati et al. 1994), and its use here facilitates comparability.

However, it is important to remember that horticultural commodities compete not just on prices, but also on several quality dimensions such as freshness, size, colour, and taste (as discussed in Chapter 2). A price-based measure of competitiveness such as the NPC thus provides only a partial picture. The quality of both domestic produce and imports is likely to vary across regional markets, as well as across time. Unfortunately, available information does not allow us to fully incorporate these dimensions into the analysis. If the average quality of imported products is markedly better than that of the domestically available produce, our NPC estimates will overestimate the competitiveness of Indian produce to the extent that the Indian consumer is willing to pay a quality premium. This is likely to be the case for a few products such as apples and grapes, though the size of quality premium is not clear. For some others such as mangoes or potatoes, however, the opposite is likely to be true so that our NPC figures might be underestimated.

Following from the earlier discussion that India is best considered a sum of regional markets, the NPCs at the commodity level for selected fruits and vegetables are estimated at four major consuming centres in India located in the metropolitan cities of Delhi, Mumbai, Kolkata, and Chennai. These cities are located in the northern, western, eastern, and southern regions of the country respectively, and are assumed to be representative of markets in these regions. They account for a majority of trade in fruits and vegetables. The all-India aggregates for each product are calculated as the weighted average of NPC estimates at these regional markets, weighted by total arrivals in each market. (For a detailed explanation of how NPCs have been estimated, see Appendix 1.)

ARE IMPORTS CHEAPER THAN INDIAN PRODUCE?

Indian horticulture is on the whole, competitive in terms of prices. We first estimate the NPC at the all-India level, aggregated over all the regions, for the period 1997–2002. This is the period when India started opening up to imports. Given the relatively short time period, the results need to be interpreted with some caution. However, this exercise confirms the findings by many other previous studies that overall fresh produce in India, fruits and vegetables, is competitive in terms of prices. Figure 3.1a shows, that between 1997 and 2002, the average NPC has consistently been below one.

The level of competitiveness varies among fruits (Figure 3.1b). In the case of bananas, domestic prices on an average are consistently less than half the price of imports when they would arrive in local markets and there is very little variation over time, that is, between 1997–2002. Grapes and oranges have slightly higher NPCs, but both appear to be becoming more competitive over time. Apples and mangoes, however, appear to be borderline competitive. It should be noted though that the last three years in the study were years of relatively

Figure 3.1: Nominal Protection Coefficients for Fruits and Vegetables: All India, 1997–2002

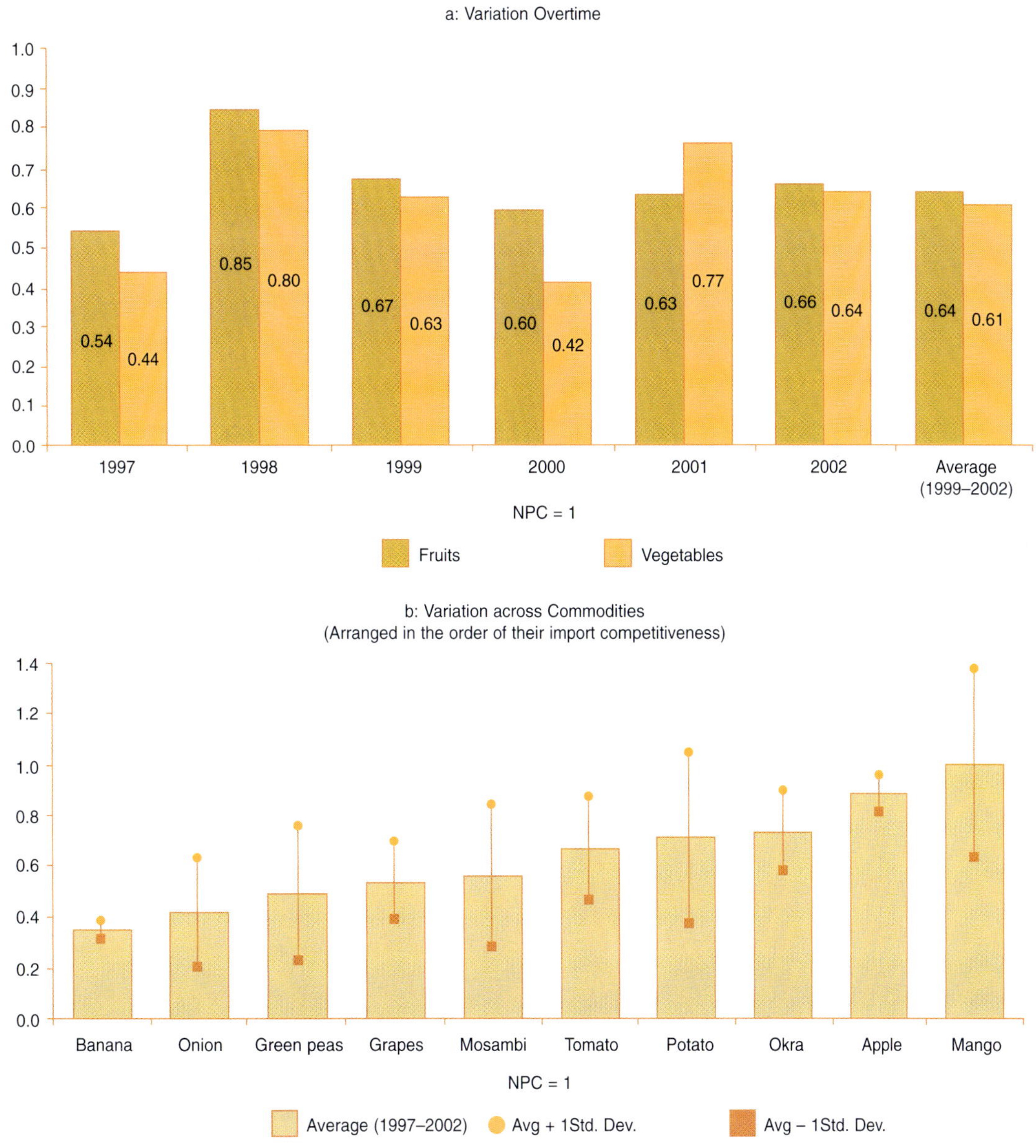

Source: NHB Survey, Staff estimates.

low productivity for apples. If productivity recovers to reach the levels witnessed in the mid-1990s, this product should become more competitive. For mangoes, the NPC was significantly higher than one only in 1998, though it has been fluctuating close to that level. The year 1998 was a bad year for mango production. An important caveat is also that comparable international prices for mango are taken as the prices prevailing in Pakistan, the nearest major exporter. Any significant purchases by India are however likely to raise Pakistani prices significantly, and erode the competitive margin estimated using current prices.

Apart from potatoes, all vegetables are also found to be competitive. Here an important caveat is that India imports very small amounts of potatoes, and that too only from its immediate neighbours (that is, Bhutan and Nepal). The unit international price used to calculate the NPCs, therefore, does not truly reflect the price India would encounter in international markets if it were to import on a larger scale. Even bearing this in mind, there is only one year, 1998, when potatoes actually became uncompetitive. This was a year when potato production was low, and prices were almost double their previous year level in all the major regional markets. Onions and tomatoes are consistently competitive. It is interesting to note that onions remained cheaper than imports even during the crisis year 1998.

THEN WHY ARE TARIFFS SO HIGH? PROTECTION OF DOMESTIC PRODUCE IN ALL REGIONS AND ALL SEASONS

India has high tariffs for even those horticulture, products that appear to be highly competitive. There could be several explanations for this. One is that the tariffs are purely precautionary, and are meant to deal with circumstances where foreign suppliers benefit from a sudden reduction in costs, thanks to an act of God or government. Another is that the market for a product is not homogeneous but differentiated, either horizontally (that is, foreign and domestic products are different varieties) or vertically (that is, foreign and domestic products are of different quality), in which case the tariff can serve a protectionist purpose. A third possibility is that tariffs are intended to ensure that the entire national market, in all regions and all seasons, is reserved only for domestic producers (even when production itself is concentrated in one region, for example, in the case of apples). A fourth possibility is that production is distributed across all regions and some of these regions are uncompetitive and the tariff is a way to protect the producers in these uncompetitive regions (this seems to be the case for crops like potatoes that are grown primarily in the poorer Indian states).

There are considerable variations in domestic prices between regions. A better understanding of the aggregate numbers can be obtained by looking at the regional picture (Figure 3.2a). On an average, prices are 70 per cent higher in the most expensive regions compared to those in the cheapest regions. In most cases, differences in price can be best explained by distances between the growing and consuming regions. For example, the major apple producing areas of the country are located in the north and north-eastern parts of India. Not surprisingly, Delhi is the cheapest market, whereas Chennai is the most expensive. Similarly in the case of green peas, Uttar Pradesh accounts for nearly half of green pea production in the country. The cheapest major market for peas is Kolkata, whereas the most expensive is Bangalore.

What is surprising, however, is that the variation in prices between different cities is 2–6 times the variation in transportation costs between these cities. This is illustrated in Figure 3.2b. The variation in prices across regions in India can be attributed to a number of reasons. First, a large part of the variation can be explained in terms of wastage that is likely to happen during transit. While road transport in India is not particularly expensive in 'nominal' terms (US$0.02–0.03/tonne-km), the problem is with the quality of the trucks and the service provided—poor and slow, which leads to high wastage. This is exacerbated by the poor or non-existent packaging of fruits and vegetables, and the lack of adequate storage infrastructure—making the 'effective' transportation cost fairly high.[1] This is illustrated in Figure 3.2c, which shows a strong negative relationship between regional price variations and the shelf life of products. Losses during transit, for example, can go up to as high as 35 per cent in peak summer (EXIM Bank of India 2002). Part of the price differential is also attributable to the inefficient distribution system where most trade is routed through just a few cities, and is mediated by a large number of intermediaries. This not only inflates prices, but also the time taken to move the product from the farmer to the ultimate consumer. Finally, some part of this price variation can be attributed to various restrictions on the inter-state movement of commodities (see discussion in Chapter 4).

There are also large variations in prices and availability across seasons. Production of many horticultural commodities is seasonal in nature. In India, seasonality effects are accentuated by the fact that storage facilities are both expensive and rare. For instance, storage capacity in the two big potato producing states of Uttar Pradesh and West Bengal was only 0.08 and 0.05 per cent of their production respectively. The problem is further compounded by poor backward linkages in the distribution chain because of which producers lack awareness about the requirements of the fresh produce industry, leading to seasonal gluts and non-availability at off-peak times. In Delhi, for example, potato prices ranged from Rs 2.91 per kg to Rs 9.53 per kg in a single year (2002). Regions also differ widely from each other in terms of availability. For example, apples are available almost throughout the year in Delhi, but are not available in Chennai markets for four months of the year. Grapes are available for eight months of the year in Delhi, but for only five in Kolkata. In general, availability of fruits and vegetables tends to be better in the north, and worse in the south. This is despite prices being higher for most products in the south.

Disaggregating the all-India figure across time and regions shows that in many cases, a high NPC in a region reflects more the effects of poor infrastructure and an inefficient distribution chain within India, rather than the un-competitiveness of domestic production. For example, the apparent price non-competitiveness of apples in the southern and western markets seems to be more the result of the difficulty that domestic producers face in serving these distant markets, than the high costs of production on part of the farmer. Overall, it appears that Indian producers face little threat of being displaced by foreign imports in markets which they can service at low cost, and protection simply means that consumers are left to pay the costs of inefficient distribution and storage.

Despite the high level of tariffs, we still see positive, and in some cases, rising imports. This could be because, as already noted, imports are better than, or different from, domestic products. Some evidence in this regard is presented in Chapter 5. Another

Figure 3.2: Regional Price Variations for Fruits and Vegetables, 2002

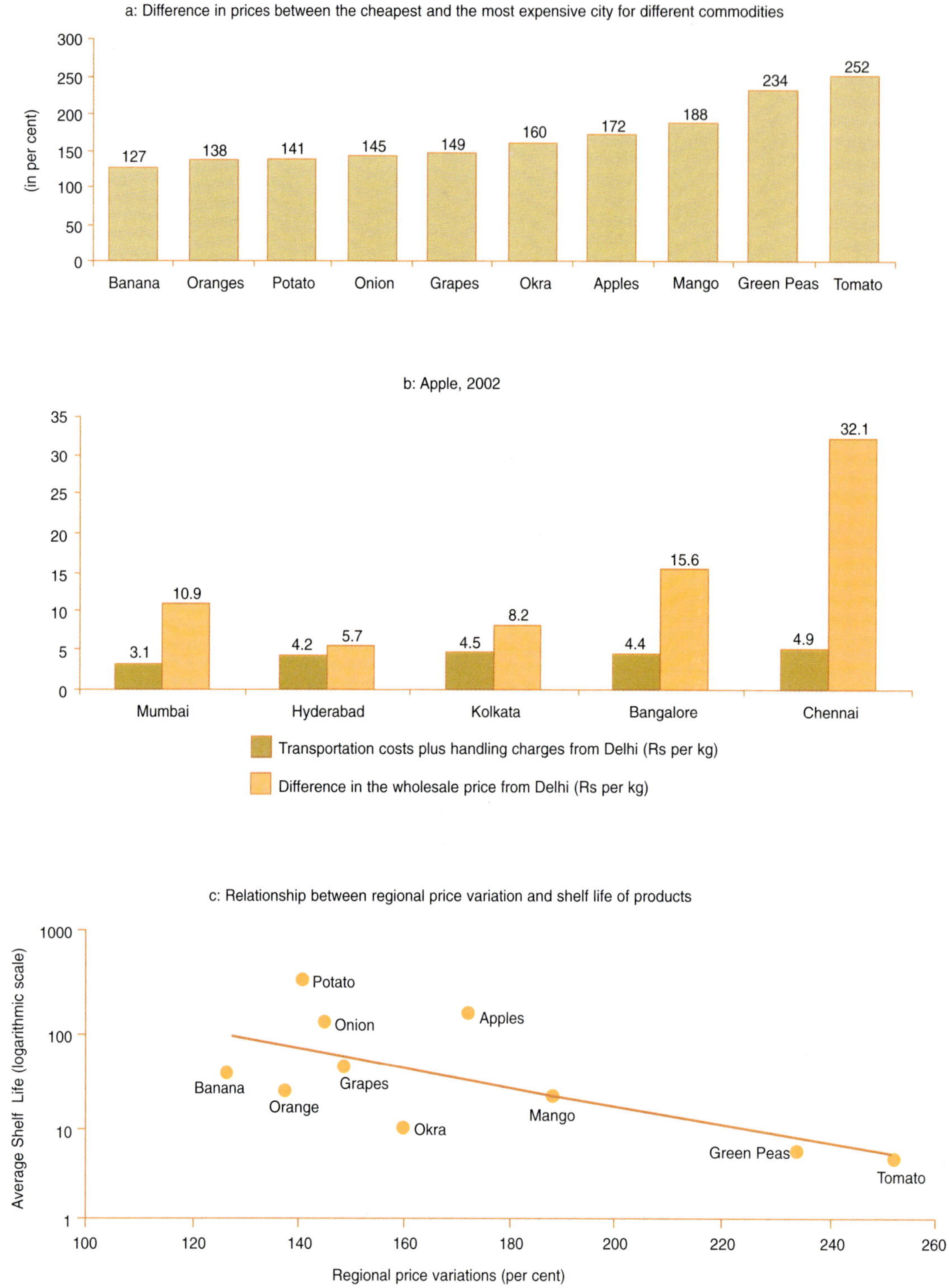

Source: NHB Survey, Staff estimates.

reason is that imports are to a limited extent complementing rather than competing with domestic production. Imports are often limited to times when domestic production cannot meet demand, either during off-peak periods or periods of domestic shortages. For example, consider the case of apples. The bulk of apple imports happen in the first half of the year, half of them just in the summer months from April–June, when Indian production is missing in most markets (Figure 3.3). Similarly imports of grapes happen mostly between August to November, which is the off-peak season for Indian produce.

To conclude this section, there is an urgent need to reconsider the tariffs applied to horticultural products. Many of the products considered in our study, such as bananas, onions, and tomatoes, were

Figure 3.3: Imports and Domestic Arrivals in India, Apple and Green Peas, 2002

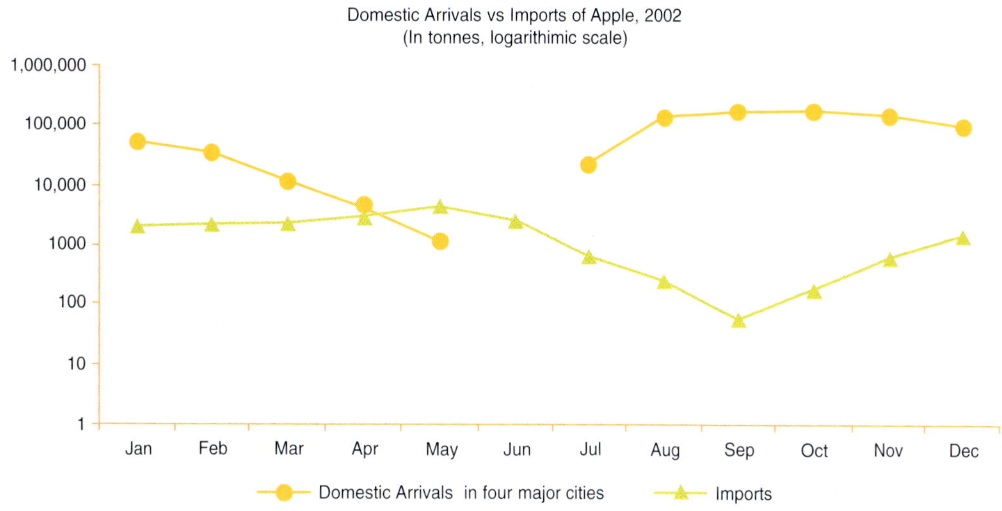

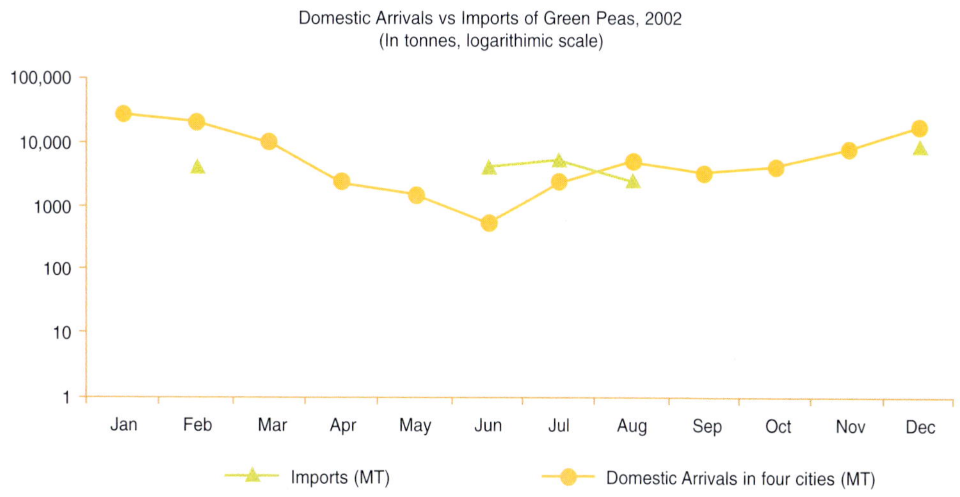

Source: National Horticulture Board Database; DGCIS

found to be unquestionably competitive across regions and seasons. These products do not require any protection, and there seems little justification and little effect of the high tariffs imposed on them—prices are largely determined by domestic supply and demand conditions. In the case of seasonal products like grapes and oranges, the low average price of Indian products compared to imports, even after factoring in the effects of inefficient logistics, means that imports are largely restricted to times when domestic producers cannot serve the market. For such commodities, tariffs raise prices for consumers, without much benefit for producers who are mostly absent from the market or present after incurring high storage costs. Eliminating tariffs for such products is likely to benefit consumers with few adverse consequences for producers. In the case of products which are borderline competitive, there is a need to assess whether tariffs are subsidizing farmers, or simply an inefficient distribution system. If, as our evidence suggests, it is primarily the latter, then the reasons for the inefficiencies assume significance—an issue that we address in the next chapter. In any case, remedying the acute inefficiencies of supporting service sectors would greatly strengthen both the political case for and economic gains from further liberalization of Indian agricture.

NOTE

1. At these distances rail should be able to provide a competitive service. However, rail rates in India are very high in relation to costs, because they subsidize all the passenger traffic, and a large part of the railway system is engaged in moving regular unit trains between coal field and power stations and/or similar activities.

4 What Undermines Comparative Advantage? The Role of Domestic Constraints

FROM COMPETING AT HOME TO COMPETING ABROAD

Since India is one of the largest and lowest cost producers of horticulture, it could be expected to be a large exporter as well. But there is a critical difference between competing at home and competing abroad—international delivery costs, especially transport costs. While high international transport costs protect Indian commodities from foreign competition at home, they are an impediment to access to foreign markets. To be a successful exporter, the country needs not only to be a low cost producer but also a low-cost and efficient transporter and distributor of commodities. In this chapter we examine the importance of 'delivery costs' in determining India's trade competitiveness and the role played by domestic factors.

In horticulture products, domestic and international transportation cost is the single-largest contributor to the retail price, accounting for nearly 25–40 per cent of the price (depending on the mode of transport). This is summarized in Figure 4.1, which tracks the journey of 1 kilogramme of grapes through the supply chain, from the farm-gate in Hyderabad to supermarkets in London by air transport. It costs between Rs 14–19 to the exporter to procure a kilogramme of grapes.[1] The costs incurred by the exporter in terms of inland transportation, storage and handling, commissions and taxes, packaging materials, plus her margin add Rs 24 to the price, resulting in an freight on board (FoB) price of Rs 43. The air-freight and the insurance charges constitute the single largest component of the retail price which adds another Rs 54, resulting in a cost, insurance, and freight (CIF) price of Rs 97.[2] So while the procurement price constitutes only 15–21 per cent of the CIF price and 11–16 per cent of the retail price, the delivery costs, which includes domestic and international costs and storage and handling charges, together account for 60 per cent of the CIF and 45 per cent of retail price. Thus purely from a trade perspective, the factors affecting the price between the farm-gate and the retail level are found to be more important than factors determining the price at the farm-gate. It is important to note that the international delivery costs are particularly high in case of air transport, which is increasingly becoming the preferred mode of transportation for international trade (Hummels 2001).[3]

We demonstrate that India's high delivery costs arise from two sources, apart from geography which we have argued in Chapter 2 is important but not decisive: (i) poor transport infrastructure and a

Figure 4.1: Evolution of Price/Cost across the Supply Chain (For a kilogramme of grapes produced in Hyderabad and sold in London)

a: Contribution of various agents of the supply chain to the final price

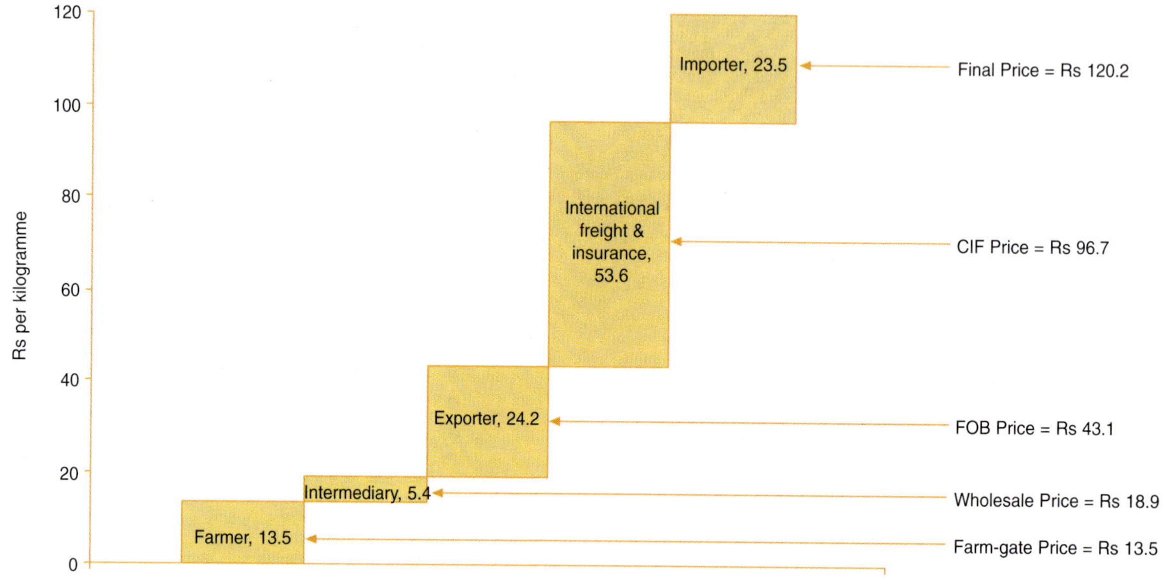

b: Contribution of various elements of the supply chain to the CIF price

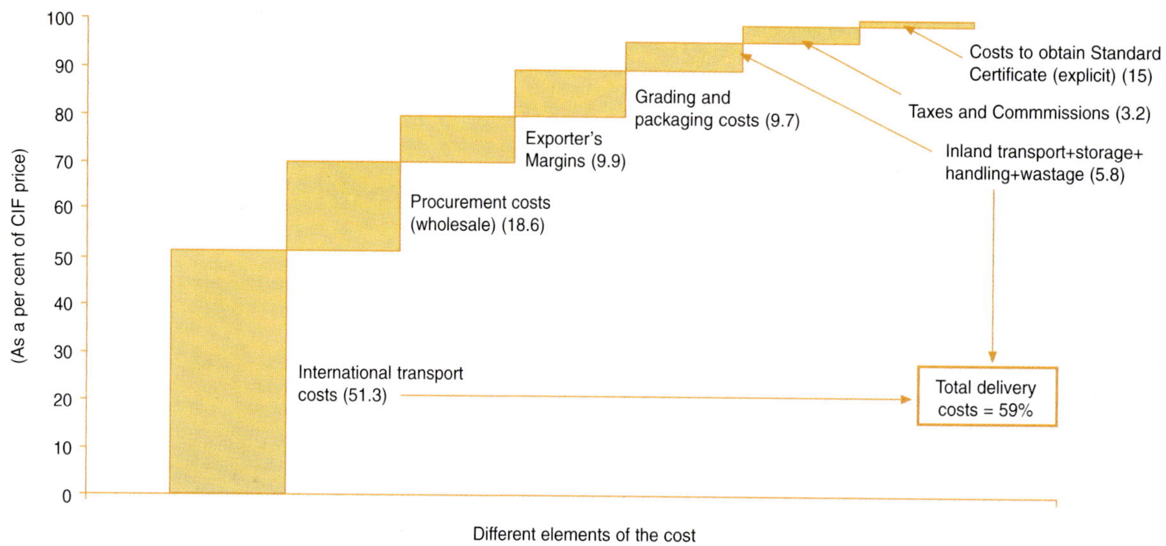

Source: Exporter survey; Staff estimates.

number of policy distortions that have restricted competition and resulted in uneven utilization of existing infrastructure and slow creation of new infrastructure; and (ii) high storage and marketing costs due to fragmentation in the supply chain and limited storage infrastructure. Inadequate consideration of these issues is perhaps the single-most important reason why many past prognoses about dramatic growth in India's food sector have not come true.[4] The rest of this chapter expands on the last two constraints (see Figure 4.2).

How inefficient is India's logistics system relative to other countries? Jaffee (2003) estimates that the cost of transporting certain vegetables (for example, bitter melon, squash, beans, and eggplants) from India to the European market is $1.50/kg, comparable with the cost of transport from Kenya but much higher than the cost of transport from other sources, such as Ghana and the Dominican Republic (from $0.60/kg to $1.00/kg). Furthermore, inefficiency in transport is reflected not only in higher explicit costs, but also in higher spoilage and handling costs that make fresh produce much more expensive to transport (Huang 2004).

Based on a number of comparisons between India and selected countries and regions in the world for different destination markets, modes of transport, and commodities traded, we find that on average, India's international transportation costs are 20–30 per cent higher than their corresponding global counterparts. Since transportation costs are found to account for nearly 50 per cent of retail price when air transport is used and 25 per cent in case of maritime transport, it would imply that Indian products could become 5 to 15 per cent more expensive than their foreign counterparts simply on account of high international transportation costs.

We, therefore, argue that a significant reduction in delivery costs is necessary for India to emerge as a globally competitive exporter of horticulture products. The current emphasis by GoI on infrastructure and a number of sector-specific initiatives in civil aviation, maritime, and road

Figure 4.2: Explaining India's High Delivery Costs

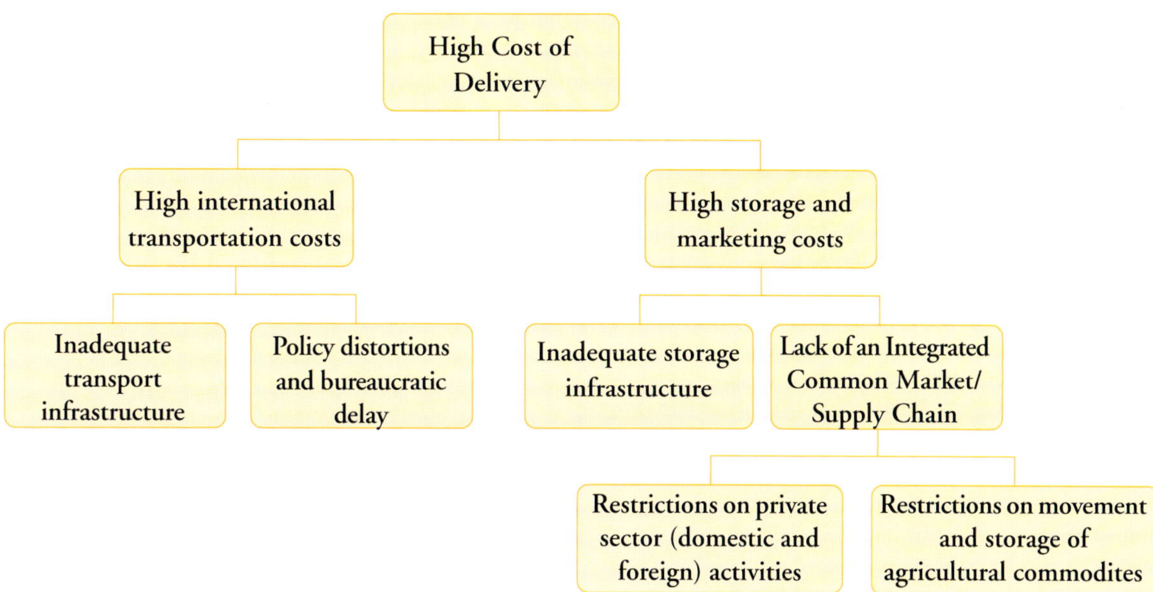

transport sectors coupled with custom reforms, if fully implemented, could help India to break the time and distance barriers as has been done by many successful horticulture exporters, for example, Chile.

TRANSPORTATION COSTS AS A BARRIER TO TRADE

High transportation cost has long been recognized as an important barrier to trade. For many countries, the protection from high transport costs has come to exceed the protection provided by tariffs. Studies have shown that countries with high-shipping costs have found it difficult to promote export-led development, even after they have reduced tariff rates, removed quantitative restrictions (QRs), and followed prudent macroeconomic policies.[5] The problem is exacerbated with highly perishable products in horticulture, which have an average shelf life of a month—making 'time' a critical factor in trade (Figure 4.3a). Given the current inefficiency in the logistic system, it is therefore not surprising that the ratio of exports to domestic production among horticulture products is strongly correlated with the degree of their perishability (Figure 4.3b).

Figure 4.3: Perishability of Horticulture Products can be a Potential Barrier to Trade

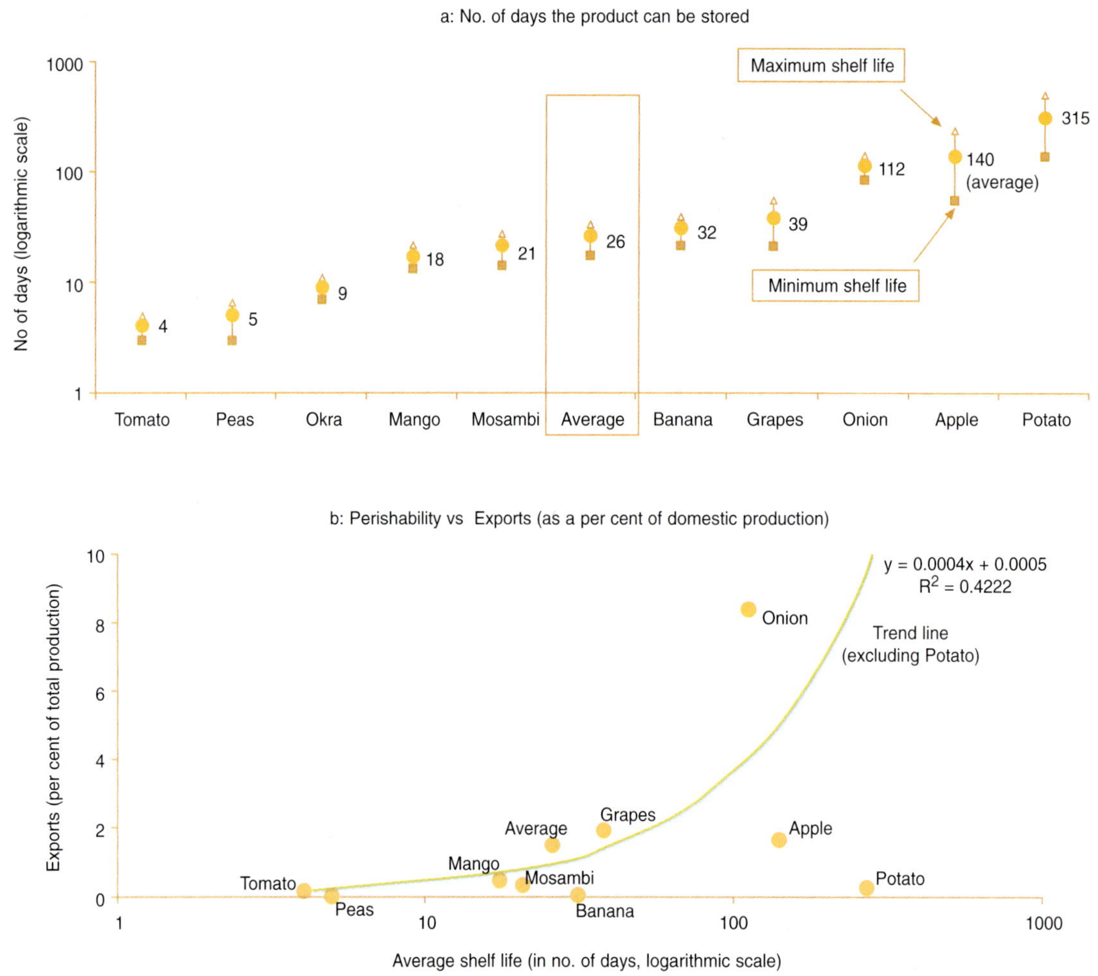

Source: UN COMTRADE Statistics, Survey of exporters (2004–5), NHB, Staff estimates.

Figure 4.4: Barriers to Trade in Horticulture

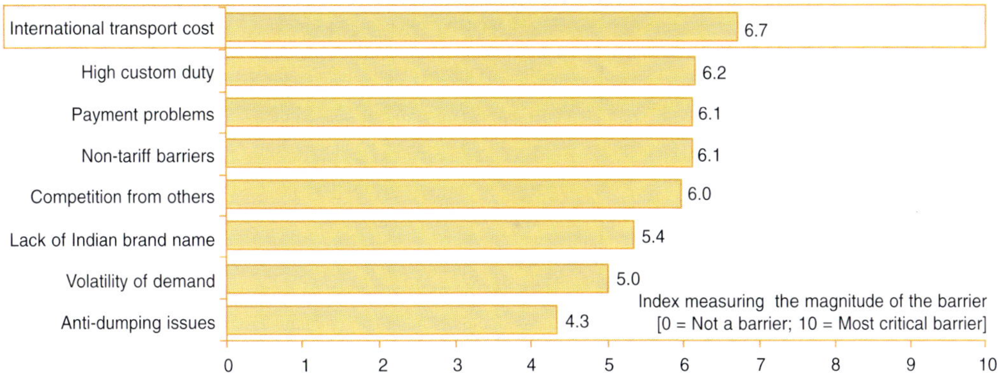

Note: Does not include behind the border domestic factors.
Source: Survey of exporters (2004–5), Staff estimates.

Holding other things constant, if the storage life of an average horticulture product can be increased by an additional 50 days, this alone could result in more than doubling of India's exports of horticulture products (the coefficient is 0.0004).

Indian exporters perceive high international transportation costs as the number one external barrier to trade. As shown in Figure 4.4, international transport costs as a barrier to trade is placed many notches above the standard tariff and non-tariff barriers (NTBs) to trade. While exporters shipping to Europe and Far East consistently placed high international transportation costs as the number one barrier, exporters to neighbouring South Asian countries viewed it as the second-most important constraint and exporters to Middle East placed it on the third spot from a potential of eight external barriers to trade. Furthermore, to an open-ended question 'What policy changes would you suggest to enhance exports of fruits and vegetables from India?' two out of every five exporters who have used air or maritime transports in the past said that government should undertake measures to reduce international transportation costs.

The concern of Indian exporters regarding international transportation costs is reinforced by the finding that it costs as much as two to three times more to transport similar products over a comparable geographical distance from India than from some of its competitor countries. This is illustrated using the example of export of grapes from India and Chile to Netherlands (Figure 4.5).[6] The distance between India and Netherlands is 6858 kilometres while from Chile to Netherlands it is 12,007 kilometres. So, based on a standard model of geography and trade, Chile's transportation costs should be around 50 per cent higher than that of India. But the actual transportation cost is found to be opposite of what is predicted by such models. It costs $790 to transport 1 metric tonne (MT) of grapes from India to Netherlands—260 per cent higher than it takes to transport the same from Chile (Figure 4.5a). Thus freight and insurance charges make up for 48 per cent of CIF value in case of India compared to only 18 per cent in case of Chile.

The above example is not an isolated observation but part of a wider observation about India's inefficient transportation system. A number of

Figure 4.5: India's International Transport Costs vis-à-vis International Standards

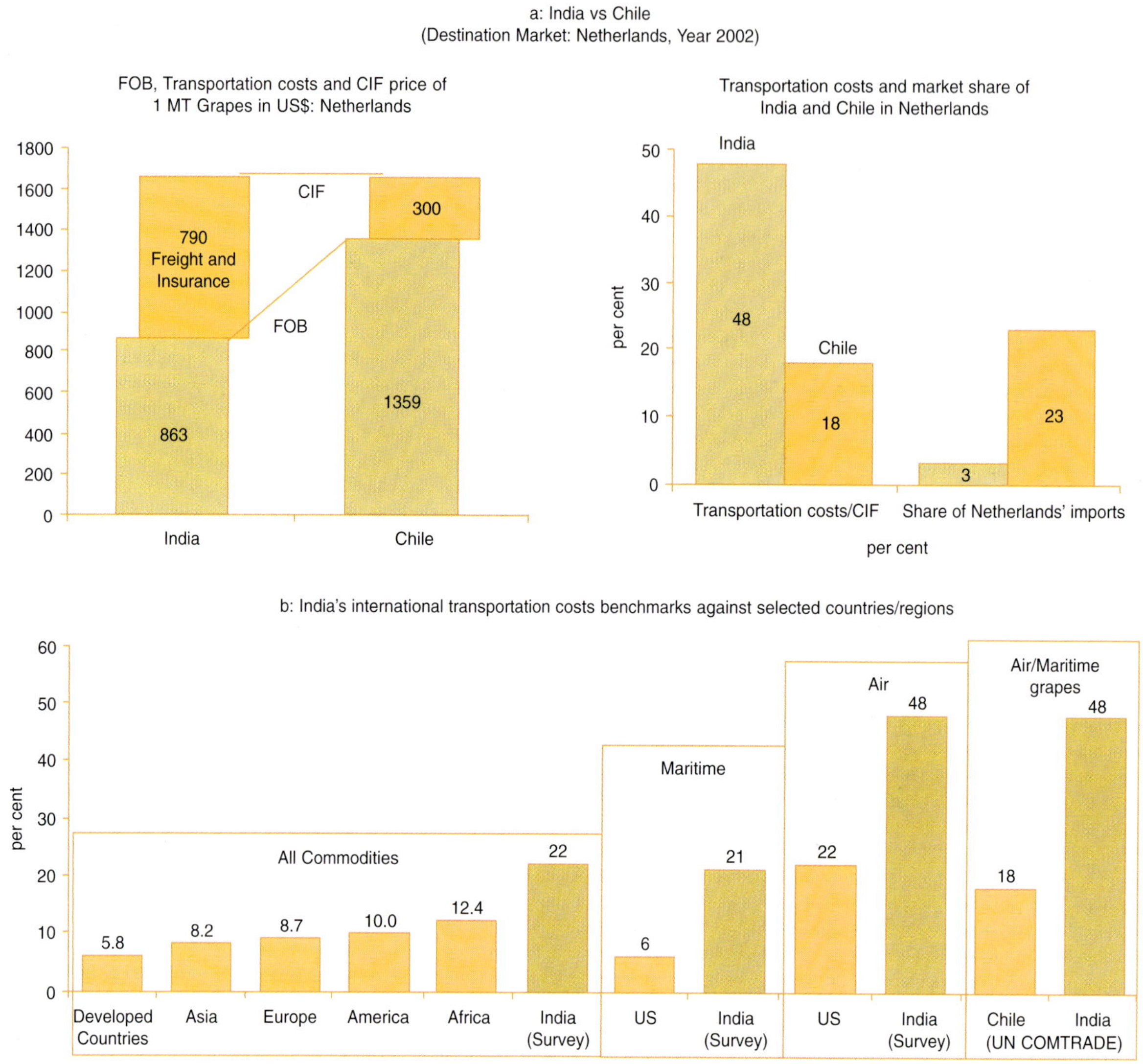

Source: UN COMTRADE Statistics, Survey of exporters (2004–5), Staff estimates.

comparisons between India and selected countries and regions in the world for different destination markets, modes of transport, and commodities traded are shown in Figure 4.5b—all of which indicate that on average, India's international transportation costs are 20–30 per cent higher than their corresponding global counterparts.[7]

The cargo dwelling time in India, which used to be considerably high in the past, has fallen in recent years, though still remains high compared to international best practice. According to Roy (2002, 2005), the time taken for clearance of international cargo in India in 1998 was five to sixteen times higher than the international norm (Figure 4.6). For example, it used to take, on average, 5 days (60 hours) to clear imported air cargoes from Delhi airport compared to the international norm of less than 6 hours. But these numbers appear to have changed in recent years. According to GoI's 'Report

Figure 4.6: Dwell Time in India vs International Norms (in hours)

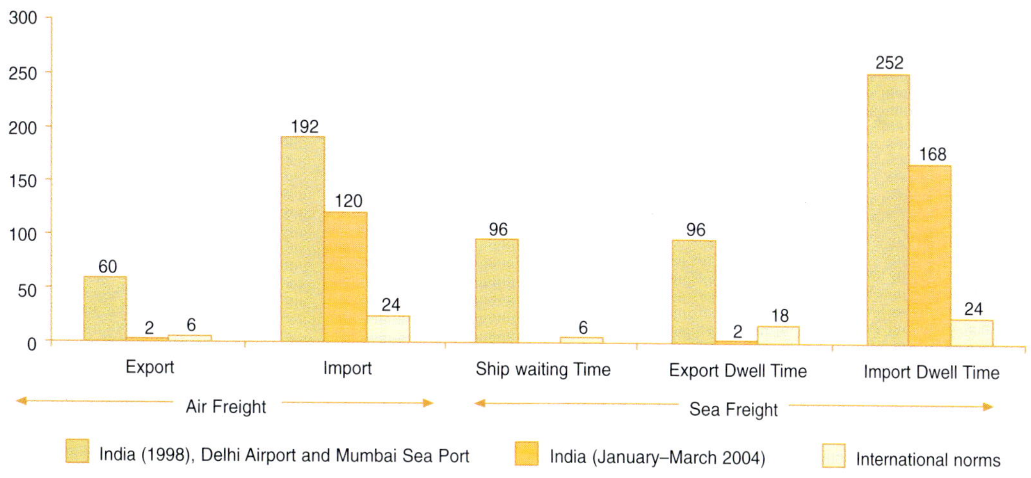

Sources: Roy (2002), Report on Working Group on Trade Facilitation (2004).

on Working Group on Trade Facilitation', the average cargo dwelling time for export has come down, though for import cargoes the situation remains considerably worse than the international norm. It is important to note that the report also states that

> dwell time figures for exports may be understated because they do not take into account the time required to get clearance from several other agencies prior to arrival of goods at customs. The process of prior clearance involves up to 257 signatures from different agencies that may take a couple of days in some cases.[8]

The high international transportation costs, it turns out, is part of the solution to the puzzle we stated at the beginning of this chapter (Figure 4.1). Indian grapes are 40 per cent cheaper than their Chilean counterparts when they leave their respective ports (in terms of FOB price) but their prices are almost the same by the time they reach Netherlands (in terms of CIF price). The delay in customs and dwelling time at ports only makes matters worse. Thus the cost advantage enjoyed by the Indian farmers is offset by its 'transportation disadvantage'. Consequently, India's export performance remains poor despite its low cost of production: while India's share in Netherlands' imports is only 3 per cent, the corresponding figure for Chile is 23 per cent.[9]

Determinants of High Transportation Costs in India[10]

While the most obvious determinant of transport cost is geography, transport cost is also considerably affected by government policies and institutions. The quality of infrastructure in air and sea-ports and roads and rail networks is likely to considerably influence transportation costs. Countries with adequate port capacity, stronger infrastructure, and more sophisticated packaging and loading technologies are likely to have lower transportation costs. The volume of trade and the predictability of shipment also seem to matter for the overall transportation cost. Moreover, port and handling fees, ease of customs clearance, and the extent of bureaucratic red tape involved also affect costs. There is now considerable agreement among Indian policymakers that its international transport infrastructure and services can be greatly improved and their costs significantly reduced.

Air Transport

India has a unilateral open skies policy for air cargo. According to this policy, any operator, including any foreign operator, can operate any number of flights by any type of aircraft to any airport having customs and immigration facilities without any bilateral agreement. The operators are also free to charge rates according to the demand and supply situation. But according to exporters, costs remain high because of low volume of trade, irregular shipment schedule, imbalanced traffic pattern (more one way freight traffic to Europe and North America), and relatively high fees and surcharges at Indian airports.

Discussions with airline representatives indicated that due to low volume of perishable exports and irregular shipment schedule, dedicated freighter planes are not yet viable on the Indian routes. A few airlines have abandoned this idea after starting operations. According to airline sources, the air transport cost could decline by as much as 15–20 per cent from current levels if the volume of exports doubles and the shipment schedules are made more regular. Fortunately, the situation is already improving, as traffic volumes are rising, and freight is increasingly becoming more two-way. Import cargo traffic handled at the international airports of Mumbai, Kolkata, Delhi, Chennai, and Trivandrum, for example, grew by an average of nearly 19 per cent per year from 2002–3 to 2004–5, while export cargo traffic grew by only 9 per cent per year over the same time.[11] Another factor is the entry of more airlines into freight carriage leading to more competition. This period the result of a series of recent government decisions. One was the signing of 'open skies' agreements with countries such as the US, UK, France, Australia, China, and Qatar. This has increased the number of international flights to be serviced by all competitors, and expanded India's linkages with global markets. Another has been the decision to limit Air India's monopoly on international flights by letting domestic carriers operate on certain international routes. However, the same time, other infrastructural constraints remain.

GoI's Committee 'A Road Map for the Civil Aviation Sector' identifies the following major problems why air transport costs are so high in India:

1. Excessively high taxes on fuel and airport charges. While fuel costs are usually 10–15 per cent of airline operating costs, in India, they are nearly 30 per cent. This is because the average price of aviation turbine fuel (ATF) in India is nearly 190 per cent higher than the international price. The excise and sales taxes constitute nearly 45 per cent of the ATF price. The fact that only government-owned oil companies enjoy exclusive privilege over supply of ATF has led to no real competition in the fuel prices. The 2005–6 Budget trimmed the import duty on jet fuel from 20 to 10 per cent, and continued the tax waiver on fuel used to service international flights. Nevertheless, local duties levied by state governments remain important, and add an average of 22 per cent to fuel costs. Furthermore, Indian airlines also cannot hedge fuel purchases through forward contracts. This reduces their ability to manage costs at a time when oil prices are volatile, as they are currently. A third factor according to the Committee report is that airport charges in India are 78 per cent higher than international averages.

2. Restrictions on foreign direct investment (FDI). At present, in the domestic air transport services, equity participation by foreign individuals/companies is capped at 40 per cent and foreign airlines are not allowed to hold equity investments

either directly or indirectly. The policy also stipulates that ownership and effective control of the airline should be vested with Indian nationals and that the airline's Chairman and two-thirds of its Directors should be Indian citizens.

a. Inadequate and underutilized infrastructure. While India may appear to have a vast airport infrastructure, only 62 of its 400 airports are currently being used. Nearly 40 per cent of passenger traffic and disproportionately large part of cargo are concentrated in the two main international airports at Delhi and Mumbai. As a result, the limited capacity at these airports has led to increased congestion, bunching of flights, and delay in passenger and cargo clearance. The situation has been exacerbated by outdated infrastructure, inadequate ground handling systems, and lack of night landing facilities. In this regard, the government's policy to invite private participation in modernizing airports over the next 5–10 years at an estimated cost of Rs 40,000 crore is welcome.

Maritime Transport

With a 5560 kilometres coastline and 12 major and 184 minor ports, India has a huge potential for growth in the maritime transport sector. Indian ports, however, handled considerably less traffic than their capacity during the Ninth Five Year Plan period, indicating that port capacity under the current trade flows is not a constraint. According to India's Tenth Five Year Plan (TFYP) report, while productivity in terms of ship turnaround time, waiting time, and average ship berth day output has slightly improved over the last decade, the performance continues to be poor when compared with generally accepted international standards and performance of regional ports. For example, according to the Global Competitiveness Report,[12] the Indian port sector operates at 35 per cent efficiency level relative to Singapore and 60 per cent relative to China. Some of this inefficiency can be attributed to superior infrastructure, for example, China has 14 times more containerized trade than India and this trade is growing at 4 times faster than India; fleet assignments to China are 8 times higher than those of India, and the Chinese national carriers have 5 times younger ships than India's (Figure 4.7). Clearly the lack of superior infrastructure has led to more time delays and higher transportation costs, resulting in lost export orders and/or higher prices of final products. Another important constraint has been that the connectivity of ports to inland markets remains weak (see Figure 4.8 for an illustration of high handling costs at the departure and destination ports). In fact, poor road and rail container evacuation facilities are widely believed to be behind the slight increases in average pre-berthing and turnaround times at the Jawaharlal Nehru Port Trust (JNPT) and 7 other of the major ports witnessed in 2004 (also see Box 4.1).

The government policy to deal with these problems has been three-fold: expanding port capacity, increasing competition, and improving connectivity. The expansion in port capacity is set to happen, relying partly on private participation. For this purpose, up to 100 per cent FDI in ports and shipping has been allowed under the automatic route, and FDI up to 51 per cent in support services such as operation and maintenance of ports and harbours, loading, and discharging vessels. There is also a ten-year tax holiday on investment in infrastructure facilities related to ports, inland ports, and inland waterways. State governments have also actively followed the Centre's lead, with Gujarat, Andhra Pradesh, Tamil Nadu, and Orissa actively seeking private participation to develop minor

Figure 4.7: An International Comparison of the Performance of India's Ports and its Maritime Infrastructure

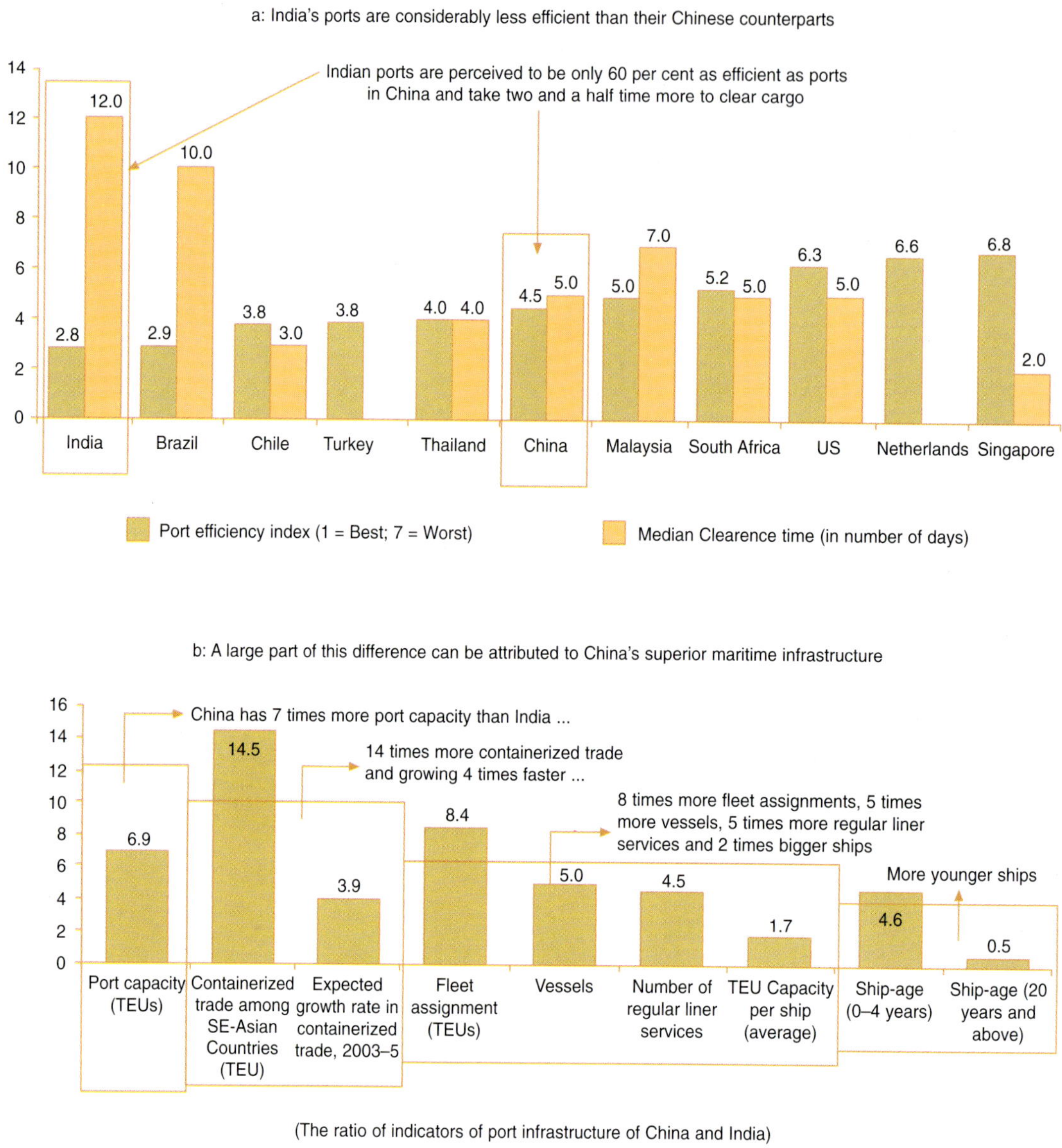

Source: UN COMTRADE Statistics, Survey of exporters (2004–5), Staff estimates.

ports. Another initative under the new National Maritime Development Policy has been to increase competition by requiring that terminal facilities at major ports or adjacent major ports be run by different operators. Steps toward improving ports' connectivity have also been taken. It will be crucial to give priority to these investments if India is to handle large forecasted increase in port traffic, especially in the container segment which accounts for most of India's high value trade (see Box 4.2).

According to India's TFYP report, the maritime sector in India faces the following challenges:

1. Bureaucratic delay has slowed implementation. The sector spent only 70 per cent of its budget

BOX 4.1: Role of Transportation Technology in Expanding Exports of Perishable Agricultural Products in the US

India's exports of fruits and vegetables could get a huge boost through the use of advance transportation technology as was seen in the US few decades ago. Perishable agricultural products, many of which US farmers could only have dreamt of selling abroad just 10 years ago, now account for about 20 per cent—a growing share—of total US food and agricultural exports. Packaging innovations, fruit and vegetable coatings, bioengineering, and other techniques that reduce deterioration of food products have helped shippers extend the marketing reach of US perishable products. In addition, new technologies in transport, a few of which are described below, are gradually opening the ocean trades to a host of perishable products. As a result, US exports of horticultural products now travel much greater distances than before.

Containerization and intermodalism. The high cost of loading and unloading is one reason transportation costs tend to decline with distance. The revolution in perishable product shipping technology began with a simple idea called containerization—handling standardized containers filled with cargo, rather than handling the cargo itself. Containerization led to a radical change in global shipping practices known as *intermodalism*—moving goods by linking two or more modes of travel. Containerization is recognized as a major contributor to the steady reduction in world transportation costs since the 1950s.

Refers. Ocean transport of cooled and frozen cargo received a substantial boost with the development of mobile refrigerated containers called 'reefers' in the 1960s. Reefers, like regular containers, are 20-foot or 40-foot boxes with their own refrigeration units. Reefers can be carried alongside general, non-chilled containers—an advantage that has challenged the competitiveness of conventional, dedicated refrigerated cargo ships that lack this flexibility. The reefer share of containerized cargo is now about 9 per cent and accounts for about 20–25 million tonnes of cargo annually. Although deep-freeze and dedicated refrigerated vessels are also important for perishable product trade—especially in carrying palletized chilled hardy fruit such as apples, peaches, pears, grapes, kiwifruit, citrus, and bananas—the reefer container trade is growing more rapidly and is considered better suited to carrying the hardy fruits as well as produce needing more careful handling.

Controlled Atmosphere (CA) Technologies. More recently, CA technologies have added refinements that have extended the shelf life of perishable products and thus expanded the types of perishables that can be shipped in reefers without spoilage. CA technologies allow operators to lower the respiration rate of produce by monitoring and adjusting oxygen, carbon dioxide, and nitrogen levels within a reefer. In this way, CA can slow ripening, retard discolouration, and maintain freshness of supersensitive perishables like lettuce, asparagus, peaches, mangoes, and avocados that would not survive well during ordinary refrigerated ocean transport. Some especially sophisticated CA systems are combined with other systems to maintain relative humidity, a crucial factor for some produce such as grapes.

Remote reefer monitoring system. The remote reefer monitoring systems can transmit and collect performance information electronically so that physical checks are not required while the reefer is stacked in the hold or on a dock. The remote system may also activate an alarm, helping minimize losses when problems arise at sea or in the container yard.

Source: Transportation Technology Eases the Journey For Perishables Going Abroad, Economic Research Service, USDA, January–February 1999.

Box 4.2: Ennore—A Successful Model for Port Privatization in India

Situated on the Coromandal Coast about 24 km north of the Chennai Port along the coastline, Ennore is the first corporatized major port of India. Originally conceived as a satellite port to the Chennai Port (primarily to handle thermal coal to meet the requirements of the Tamil Nadu Electricity Board), Ennore port's scope was expanded on account of subsequent developments such as the state government's plan to set up a 1880 MW LNG power project with private collaboration, a large 'Petro Chem Park', and a Naphtha Cracker Plant. Pollution and environmental hazards caused while handling iron ore and coal in the Chennai harbour and nearby habitations also necessitated shifting these cargo items from the Chennai Port.

Organizational model. Ennore is organized as a 'Landlord Port',[14] where the port authority retains the port infrastructure and regulatory functions while port services are provided by private operators. The port authority constitutes a landlord managing the basic port assets by letting land and infrastructure to port operators in an efficient manner. It is involved in planning, lease negotiation, safety, navigation, and overall co-ordinating functions. Cargo services, marine service, ancillary services, berths etc. are privatized on a captive/build, operate, transfer (BOT) basis to the primary port users. Port operators and other undertakings that need to be located in the port, lease the land, infrastructure, and associated services and provide them to the secondary users—cargo owners, ship owners, and cargo ship owners.

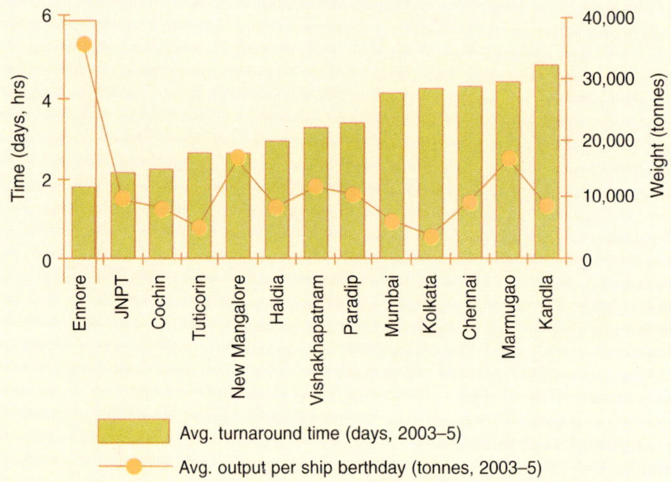

Salient features. This organizational model has enabled Ennore to develop into a one-stop shop Logistics Services Provider. It boasts of an all- weather port, 24-hour operations, a strategic location positioned within India's IT-driven integrated logistics chain, vast back-up land, simple documentation requirements, an eco-friendly environment, state-of-the art infrastructure, modern navigational facilities, efficient logistics systems, excellent transport interface, access to quality human resources, and a customer friendly approach. It also has separate berths for users of coal, iron ore, LNG, POL, chemical and other liquids, and crude oil, contributing to its evolution as a multi-functional energy hub for Asia.

Outcome. All this has translated into Ennore being the choice destination for liquid bulk cargo and the most efficient major port in India, based on operational shipping parameters (see figure). In fact, its efficiency has improved even further since 2003. Given its impressive facilities and performance, Ennore clearly serves as a model for privatization of ports in India and an illustration of successful reforms in this sector.

Source: Ministry of Shipping.

allocation during the ninth plan period due to delays in sanctioning of schemes, slow progress of work by contractors, contractual disputes/litigations, delays in tender finalization, award of contracts, deferment of projects/schemes, and weeding out of some schemes.

2. Functioning of major ports. The functioning of major ports under various port trusts, according to the report, is operationally inflexible and they are unable to respond quickly to changing market situations due to delays inherent in the decision making process.

3. Lack of multi-modal transport. India is yet to have a mainland gateway port. At present, about 70 per cent containers are trans-shipped at Colombo, Dubai, or Singapore increasing transit time and adding to cost. The idea to develop JNPT and Chennai ports, which are capable of berthing mother vessels, as mainland gateway ports connected by a rail/road bridge and equipped with efficient and modern container handling facilities has not made much progress.[13]

4. Tariff policy and regulatory authority. The tariff structure currently is determined by cost-plus approach and thus the inefficiencies in the system are being absorbed by the users of the maritime services such as the exporters of horticulture products.

Surface Transport

As indicated in Chapter 3, the nominal road transport cost in India is not particularly high, that is, \$0.02–0.04/tonne-km. But the effective road transport costs, which include wastage during transportation, loading and unloading charges, and storage costs, can be as high as or more than the nominal cost of transportation. This is because the quality of trucks and the services provided are poor and slow. Only small portions of the major corridors are dual carriageways (although this is changing fast), and few of these are outside the large cities. Many of India's major roads are poorly maintained and congested. The result is low average travel speeds, in the range of 200 to 400 kilometres per day compared to 800 to 1000 kilometres in developed countries. The movement of containers on the Indian roadways is limited not only by the design and condition of the roads and traffic congestion but also by non-physical barriers to moving containers out of the port. Tractor-tailers are also rare because of their cost, as well as road conditions, congestions, and weight limits. Moreover, trucks with refrigeration facility are few and the cost of refrigerated transport is two and a half to three times more than transportating the same by ordinary trucks. But the road transport sector is one where there has been noticeable improvement in recent years (for example, Golden Quadrilateral, East–West, and North–South road networks) and a number of new initiatives are in the pipeline. The approach paper to the Eleventh Five Year Plan envisages a quadrupling of road construction activity over that achieved in the Tenth Plan, which in itself was a big jump.

Given the vast size of India, rail transport should be able to provide competitive service but rail rates in India are very high relative to costs. Despite recent efforts to improve performance, railways in India continue to suffer from overstaffing, poor maintenance, old rolling stock; and cargo rates are particularly high due to cross-subsidization of passenger traffics. Rail share in total cargo traffic,

Figure 4.8: International Transport Cost Incurred at Departure and Destination Ports

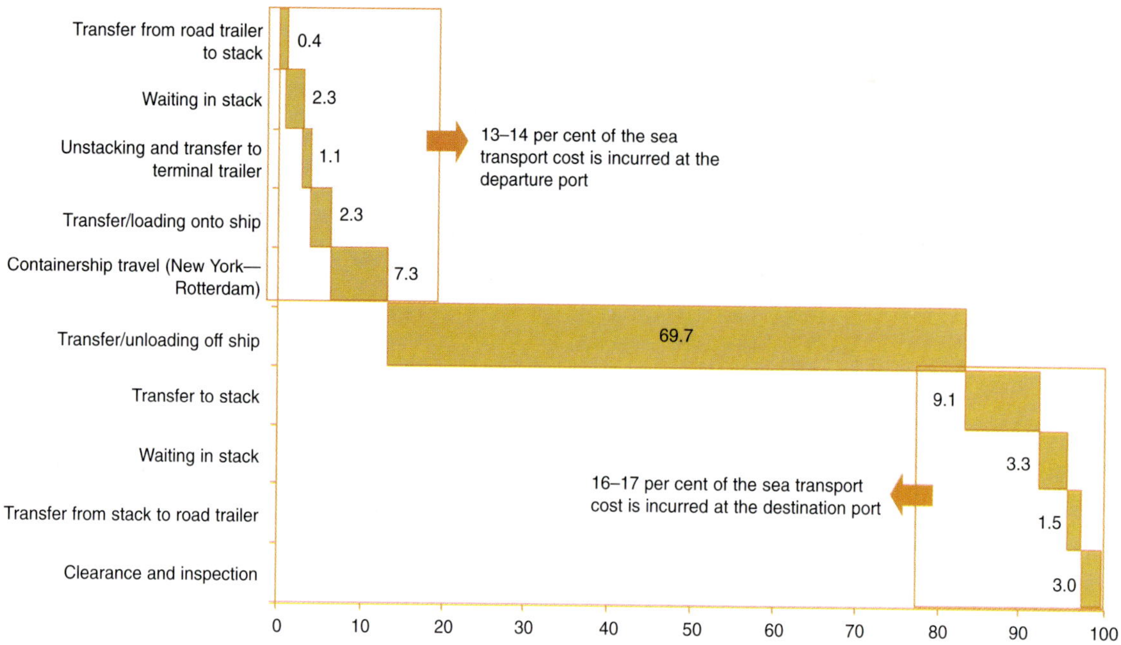

Source: Transportation Technology Eases the Journey For Perishables Going Abroad, Economic Research Service, USDA, January–February 1999.

which was around 50 per cent in the early 1980s in terms of tonne-kilometres, dropped below 40 per cent by 1992 and continued to decline thereafter, although tonnage has recently increased. Given its public sector nature, Indian railways have also tended to cater mostly to the need of public sector enterprises and train movements are, therefore, more common between coal fields, oil refineries, power stations, and oil distribution points. However, after the formation of the Container Corporation of India and the procurement of a large fleet of cars of transporting standardized boxes, the movement of containers by railroad has increased substantially.

At the same time, the railways infrastructure is still extremely congested, particularly in areas most important for freight. Only a quarter of the rail network is double track, so that outdated signalling equipment and traffic management create a significant bottleneck. Freight trains run at about half of the speed at which modern locomotives should run.[15] It also faces a shortage of wagons to carry containerized cargo on which much of horticulture trade depends. Despite substantial additions to its stock, CONCOR, the container shipping arm of Indian Railways, was not able to keep up with demand. Consequently, the wait time for importers and exporters to move goods in and out of ports was at least a week.

To ease these constraints, Indian Railways has proposed an ambitious programme of capacity enhancements, paralleling that underway in the highways sector. It has proposed investments worth $100 billion over the Eleventh Plan period, nearly 60 per cent of which are seen as coming from the private sector. To ease the shortages of wagons, the government has recently (2006) taken a decision to end CONCOR monopoly on freight transportation by allowing the entry of private players. Under the new scheme, while the railways will continue to provide engines and drivers, procurement of rolling stock will be the responsibility of the private provider. This is expected to bring new investment into the area, and ease shortages. Private players have also been given the freedom to decide their tariffs for rail haulage, terminal handling, and ground rent. Other measures are aimed at improving the network. Proposed projects include shifting from narrow to broad gauge for railways to make sure railroads can connect to main trunk lines. Very recently work has started on a dedicated freight corridor to ensure hassle-free cargo transit between various ports on the western coast and hinterlands in Delhi, Punjab, Haryana, and UP. This is aimed primarily at container traffic. Another corridor starting in Ludhiana in Punjab, and connecting steel plants in east India to the proposed deep sea port in Kolkata is also on the anvil. The two corridors will be connected to facilitate inter-transfer.

A recent World Bank report[16] observed that 'interregional' (that is, within South Asia) shipments of fruits, vegetables, and other perishables from Bhutan and Nepal to India and Bangladesh require much better logistics. This can be achieved by allowing the cargo to move in a single truck from origin to destination and by ensuring that the clearance time at the border on both sides does not exceed six hours.

IMPACT OF STORAGE AND MARKETING INFRASTRUCTURE

The single-most important indicator of the lack of adequate storage and marketing infrastructure for horticulture is the level of wastage in the system—which according to some estimates is as high as 20–40 per cent of the total production.[17] Our survey results indicate that the average wastage in the horticulture sector, between the farm-gate and the wholesale level, is around 12 per cent of the total production (Figure 4.9)—which is 800 per cent of India's total horticulture exports (in terms of quantity).[18] In fact, according to one study, India wastes more fruits and vegetables than those consumed in the UK. The level of wastage is found to be higher for more perishable commodities like tomatoes, bananas, and grapes—in these commodities farmers tend to lose 10 to 15 per cent of their produce while transporting them from the

Figure 4.9: Wastage at Various Stages of the Value Chain (per cent of total production)

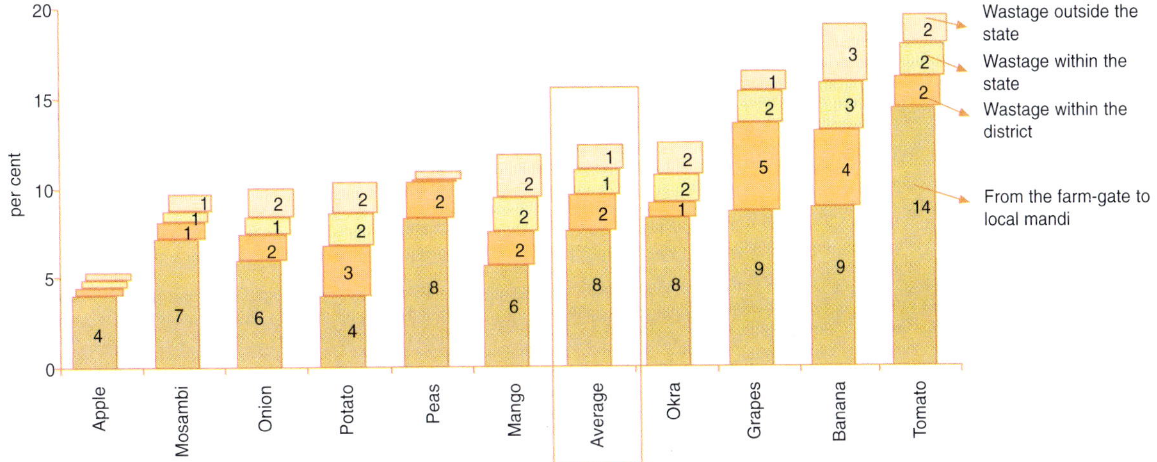

Source: Survey of Exporters, Commission Agents and Growers (2004–5), Staff estimates.

farm-gate to the local *mandi* or to the nearest trans-shipment point. The wastages at the mandis, and during transportation of the commodity between mandis (or to the ports), are also high.

The high level of wastage in the system is due to poor and multiple handling, improper bagging without crating, lack of temperature-controlled vehicles and storage facilities, and poor infrastructure (roads, warehouses, and market yards). One of the important links in this food chain is the regulated market (organized mandis),[19] which is the first stop for the commodity after it leaves the farm-gate. Given that the majority of horticulture produce around the country reaches a market yard within 4 to 8 hours of harvest, the level of wastage and the quality of produce is expected to be significantly affected by the level and quality of infrastructure in these yards. In our survey, we asked farmers about the availability of infrastructure (and management practices) in their local market yard and how satisfied they are with the existing set-up (Figure 4.10).

The level of key infrastructure in these market yards (also called mandis) is highly inadequate, but where they exist, they seem to be properly utilized. Only

Figure 4.10: Infrastructure in Local Mandis is Perceived by the Farmers to be Inadequate and of Poor Quality

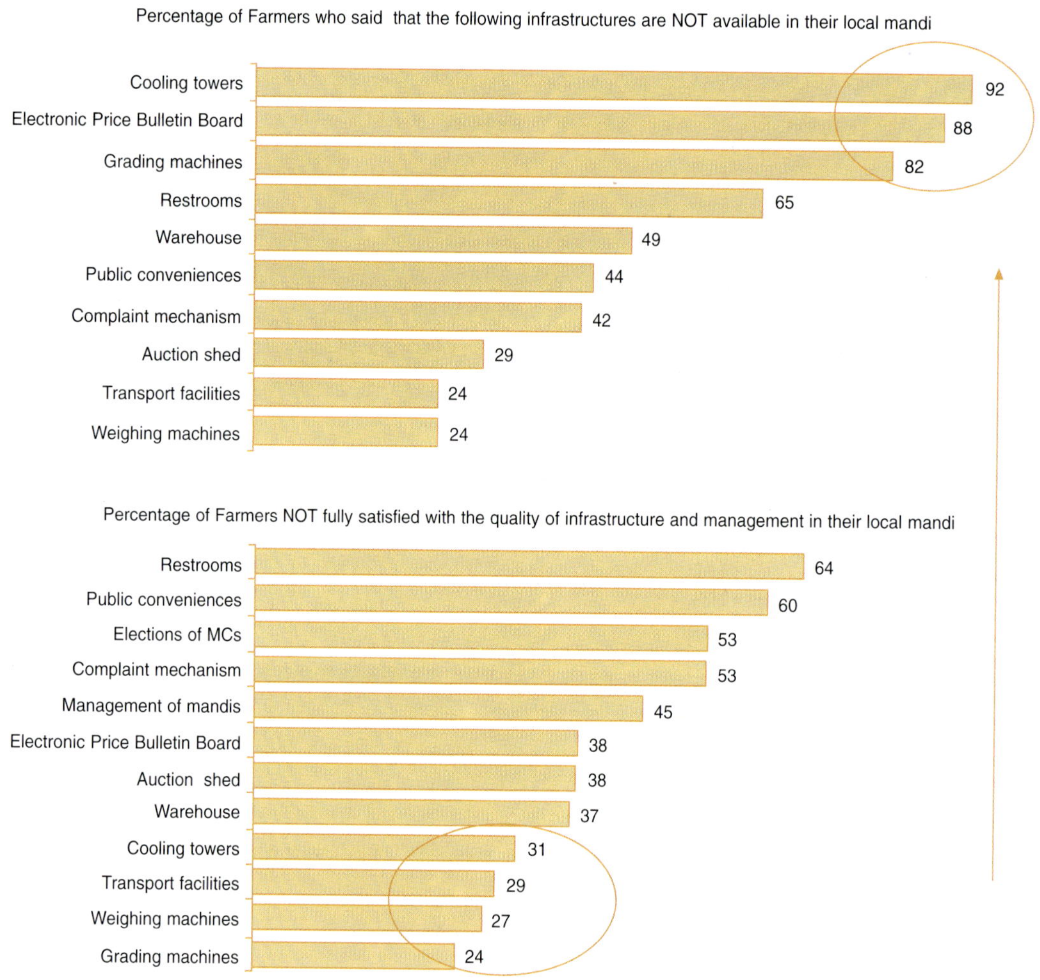

Source: Survey of growers (2004–5), Staff estimates.

one out of ten mandis had a cooling tower or electronic price bulletin board and only two out of ten mandis had a grading machine. Even basic infrastructure such as weighing machine was unavailable in two of the ten mandis, as reported by the farmers. In mandis where key infrastructure exists, a large number of farmers seemed satisfied with the quality of their usage—for example, two to three out of ten farmers interviewed felt that cooling towers, grading machine, and weighing machine, if they exist, seem to operate to their satisfaction. In the survey of 1600 growers, we find that those who did transact in mandis with relatively better infrastructure also received a better return on their produce.[20] In fact, the presence of better infrastructure in mandis is found to make almost Re 1 per kg difference to price received by the farmers (which is 20–25 per cent higher than prices obtained by farmers who do business in a mandi with average infrastructure).

Not only is the absolute level of infrastructure in the regulated markets quite inadequate, there is also considerable disparity between Indian states (Figure 4.11). The distribution of current infrastructure

Figure 4.11: Considerable Variation in the Amount and Quality of Infrastructure Available in the Local Mandis across States

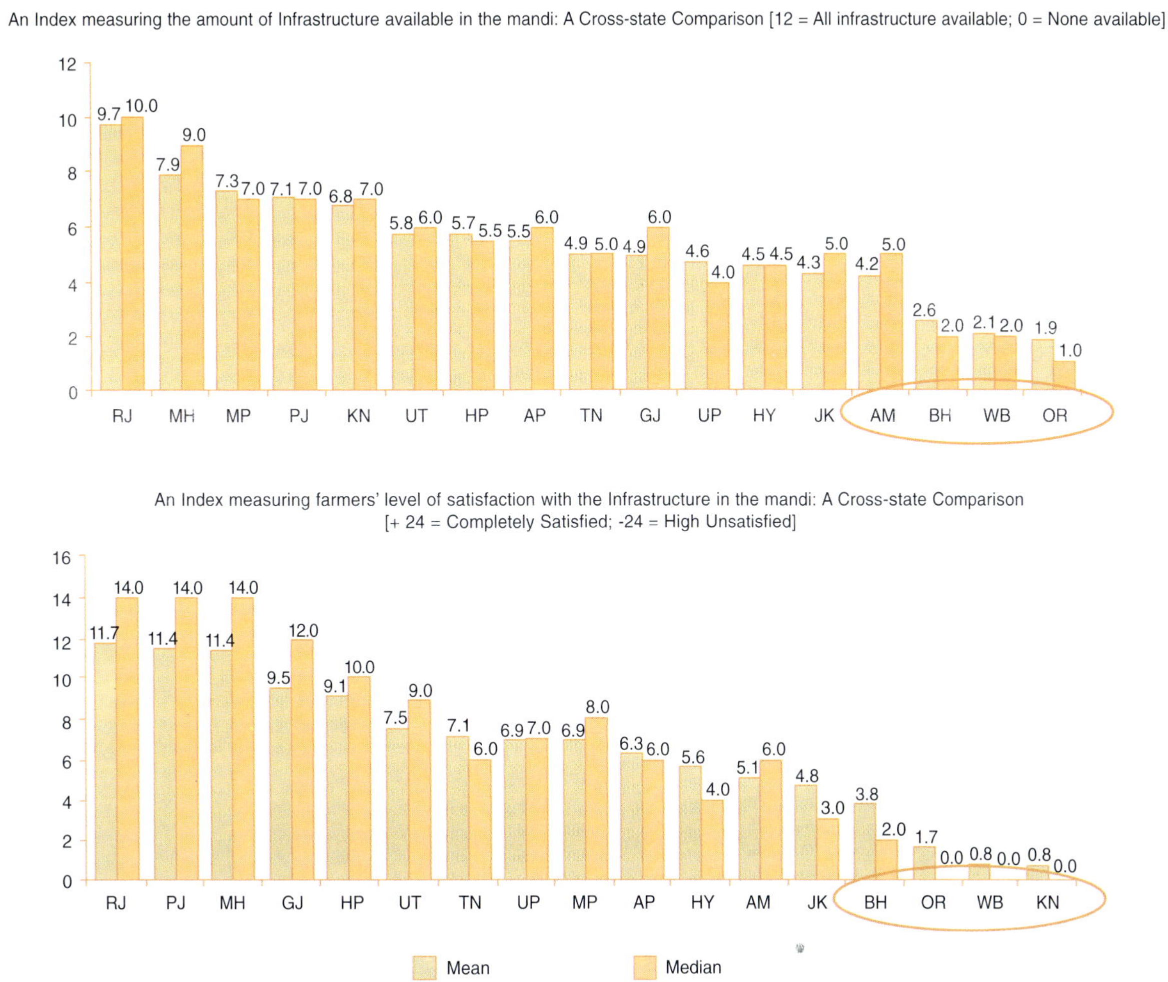

Source: Survey of growers (2004–5), Staff estimates.

seems quite regressive, that is, richer states have better marketing infrastructure than the poorer states (Rajasthan and MP are the exceptions to this trend). This is disconcerting, because farmers in the poorer states are likely to be rendered less competitive due to poorer infrastructure and thus less likely to be engaged in exports of horticulture products compared to their counterparts in the richer states.

LACK OF AN INTEGRATED COMMON MARKET AND FRAGMENTED SUPPLY CHAIN

Production in horticulture is scale neutral but storing, marketing, and distribution (wholesale and retail) are subject to economies of scale. A number of government policies have prevented the realization of these economies. To increase its control over storage, marketing, and distribution of agricultural commodities during periods of food shortages within the country, the government enacted the ECA (1955), APMA (1972), and the Prevention of Black Marketing and Maintenance of Supplies of Essential Commodities Act (1980) (see Box 4.3). However, these policies had two unintended negative effects. First, since agriculture is a state subject, many Indian states introduced legislations similar to the one adopted by GoI, preventing free mobility of agricultural produced and thus segmented the Indian domestic market into many smaller markets. Thus the supply chains that developed under these legislations were primarily local or regional in nature, which started and ended within the same state boundaries. This, along with restrictions on private domestic investment in wholesale and foreign investment in both wholesale and retailing of agricultural commodities, prevented the emergence of a large, organized, efficient pan-Indian supply chain in the agricultural sector.

Second, the surfeit of restrictions on the movement and storage of commodities (including reservation of cold storage and agro-processing industries for the small scale sector) discouraged and/or prevented the private sector from undertaking large-scale investment in agricultural storage and marketing activities. With state finances under stress, the public sector could not undertake the required investments either (Figure 4.12). Consequently, India's supply chain became populated with too many small and inefficient intermediaries, who (and not the farmers) have gradually become the largest beneficiaries and the biggest supporters of the excessive regulations in the agricultural storage and marketing sphere.

In India, marketing chains are highly fragmented, often with six to eight intermediaries, and are dominated by small-scale enterprises. At each stage of the chain, not only one incremental costs and margins built into the price, but even the time in hauling commodities through the chain increases. Some of the most complex and inefficient procurement systems in India are structured as follows: The first link in the chain is a consolidator at the village level who collects the marketable surplus from some of the farmers. The consolidator in turn sells the produce to a commission agent. The agent passes on the produce to a trader who operates at the district level and who then supplies fruits and vegetables to various large wholesalers. The wholesaler, having made purchases from a number of sources, then breaks the bulk and supplies the produce to individual retailers. Because markets are inefficient, farmers tend to receive a small share of the (domestic) consumer price—only about 25

BOX 4.3: Regulatory Policies Affecting Domestic Trade in Agriculture

ECA, 1955. The ECA aims at ensuring the availability of reasonable prices of essential mass consumption items, including foodstuff, by providing the GoI with considerable powers to control and regulate the production, supply, and distribution of these commodities. While GoI has temporarily lifted several key regulations of the ECA, it has not repealed it, nor have the state governments.

APMA, 1972. The APMA puts restrictions on wholesale transactions involving agricultural goods including horticultural products. According to the APMA, all wholesale purchases in any given place should take place in the government notified market yards. And these market yards are to be built, maintained, and regulated directly by the market committees, which comprise representatives from both public and private sectors. This Act is now being amended. A GoI Taskforce has completed the Model Act, which eliminates the exclusive state-regulations and permits the private sector to undertake development of wholesale markets. Several states including Andhra Pradesh, Maharashtra, and Punjab have indicated their commitment to implement GoI's model APMA.

The Prevention of Black Marketing and Maintenance of Supplies of Essential Commodities Act, 1980. This Act empowers senior government officials to arrest any person whom he views as preventing him from acting in any manner prejudicial to the maintenance of supply of essential commodities. It does not entitle the arrested person to appear by any legal practitioner and the opinion of the Advisory Board which reviews the case is kept confidential.

Small Scale Industry (SSI) Reservation, 1977. The SSI reservation policy reserved 3500 items for exclusive production by small-scale and cottage industries. Among the items that were included in the SSI list are 'cold storage' and 'agro-processing' industries, including those that prepare preserved and processed fruits and vegetables. Both these items have however been recently de-reserved.

Source: World Bank (1997, 2004).

Figure 4.12: Secular Decline in Investment in Indian Agriculture

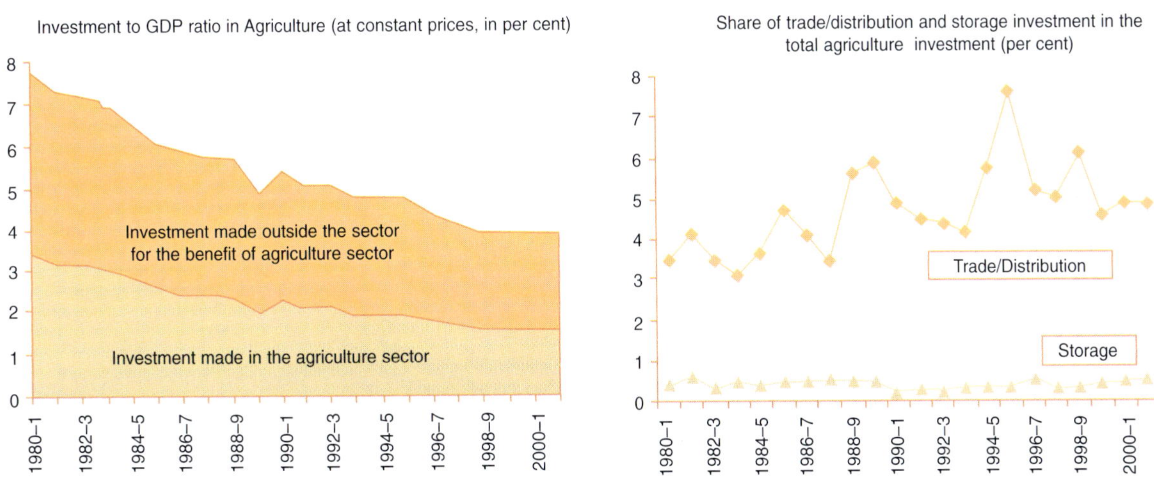

Source: Task Force on Capital Formation (GoI), Staff estimates.

per cent in the case of unprocessed vegetables compared to 40 per cent in developed economies. Physical losses in the food chain, as we have indicated earlier, are high as well—12–40 per cent for horticultural products.

A Comparison of Agricultural Supply Chain: India vs the US

Unlike in India, the fresh produce market in the US has changed markedly over the last fifteen years. Shifts in consumer demand, technological changes in production and marketing, and retail consolidation have altered the traditional market relationships between producers, wholesalers, and retailers.

In the US, fresh fruit and vegetable products move quickly through the marketing system to combat spoilage. After harvest, fresh produce is handled and packed either by a shipper or grower-shipper. Produce grown in the United States may be exported, or sold direct to consumers, retail stores, or foodservice establishments. Sales from grower-shippers to retailers and foodservice establishments might be mediated by wholesalers or brokers, or might occur directly (see Figure 4.13). Prior to 1987, fresh fruit and vegetable markets were more fragmented; most transactions took place between produce grower-shippers and wholesalers on a day-to-day basis, based on fluctuating market prices and quality levels. Today, a typical produce sale may take place between a multi-product grower-shipper and a large supermarket retailer under a standing agreement or contract specifying various conditions and terms, including marketing services provided by the grower-shipper, volume discounts, and other price adjustments and quality specifications. Some of the key trends in the fresh fruit and vegetable market in the US in recent years include the following.

Consumers

American consumers are spending more on fresh produce, and in addition to buying a greater quantity of produce, they are buying new value-added products. Per capita consumption of fresh fruits and vegetables increased 6 per cent between 1987 and 1995, and 8 per cent between 1995 and 2000. As consumption has increased, so has the demand for variety, convenience, and quality—the variety of fresh produce items carried by retailers has increased from 173 items in 1987 to 354 items in 1997. Many products (for example, lettuce and tomatoes) are available year-round, produce is pre-cut, and more packaged and branded products are available. The share of branded produce increased from 7 per cent in 1987 to 19 per cent in 1997, while fresh-cut produce and packaged salads rose from 1 per cent to 15 per cent of total sales. At the same time, the share of income spent on food-at-home purchases continues to fall. Consumers spent almost 47 per cent of their food dollars in the foodservice/restaurant sector in 2000, compared with 44.7 per cent in 1987 and 46.6 per cent in 1997.

Growers

While a typical Indian farmer producing fresh fruits and vegetables tends to sell almost all of her produce to the local consolidator (Figure 4.13a), the grower in US has several options to consider, for example, direct market, general-line grocer wholesaler, specialized producer wholesaler, general-line foodservice wholesalers, retail store, and foodservice establishments (Figure 4.13b). While

Figure 4.13: A Typical Supply Chain of Fruits and Vegetables in India vs More Mature Markets (for example, US)

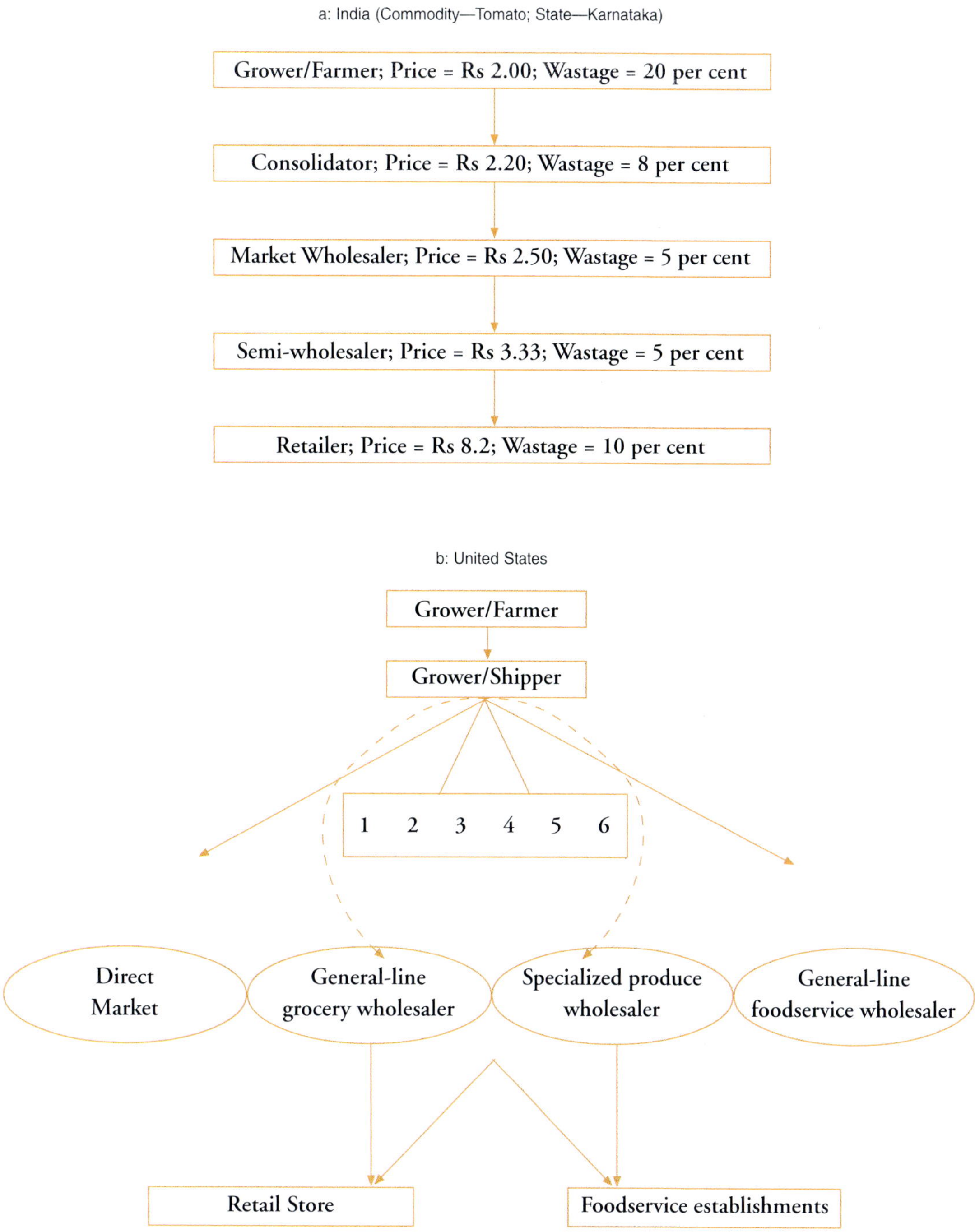

Source: a: Raghunath and Ashok (2004).
Source: b: *US Fresh Produce Markets Marketing Channels, Trade Practices, and Retail Pricing Behavior*, By Carolyn Dimitri, Abebayehu Tegene, and Phil R. Kaufman, Agriculture Economic Report #825, September 2003.

the growers in US tend to receive a higher share of the retail price (around 30–40 per cent depending on the nature of produce) than their Indian counterparts (whose share tends to be between 15–25 per cent), but with growing consolidation of the market at the shippers and retailer ends, the market power of growers has also altered over time.

Shippers

Compared to the Indian market, the shipping industry in the US is highly concentrated. For example, in 1999 there were 149 California grape shippers, 25 California tomato shippers (down from 31 in 1996), and 54 bagged salad firms selling to retailers (down from 63 in 1994), with the top two firms accounting for 76 per cent of total fresh-cut salad sales in supermarkets. In fact, for a few fresh produce items, the concentration of sales at the shipper level is as much or more than that of the retailers.

Wholesalers

Anecdotal evidence suggests that the role of merchant wholesalers in produce has diminished while that of the foodservice sector (examples, restaurants, offices, and schools with cafeterias) has grown. Large retail stores have increased the volume of direct purchases (from grower-shippers), bypassing produce wholesalers. But at the same time, wholesalers have increased their share of produce sales to the foodservice channel—from 8.4 per cent of sales in 1987 to 21.2 per cent of sales in 1997.

Retailers

In the US, the emergence of supermarkets has had the largest impact on the way the wholesalers, grower-shippers, and other intermediaries of the fresh fruits and vegetables supply chain operate and this experience is expected to get repeated in India as new investment into organized retailing takes hold. In the US, retail consolidation has occurred rapidly as large supermarket firms have merged or have been acquired. Mass merchandisers such as Wal-Mart, Kmart, and Target and warehouse club retailers like Sam's Club are selling an increasing volume of food products with low-price strategies. They have also introduced innovations in the procurement and distribution of the products they sell to gain efficiencies and lower costs (Kinsey 2000). Some mass merchandisers, such as Wal-Mart, provide suppliers with real-time store sales data to support distribution, inventory management, and in-store promotion activities (Kinsey 2000). Wal-Mart also introduced a standardized returnable (to the supplier) plastic container that is used both for distributing fresh fruit and vegetable products to their stores and for in-store product display. The Wal-Mart model emphasizes co-operation and co-ordination of activities in the supply chain between suppliers and buyers, with the goal of reducing system-wide costs. Supply-chain management practices such as continuous inventory replenishment, the use of cross-docking facilities, direct store delivery by suppliers, and selective use of speciality wholesalers has reduced the need for large distribution centres and their associated costs. To achieve these efficiencies, retailers are consolidating, as evidenced by a significant increase in the share of total US grocery store sales by the largest firms. By 2000, the share of the 20 largest retailers had reached 52 per cent of total grocery sales, up from 36.5 per cent in 1987.

Mirroring the trend in the US, the global supply chain for agriculture, especially horticulture, has changed markedly in recent years with the emergence of supermarkets as a major buyer force. There is a much higher degree of vertical integration to meet the increasingly high standards. Compliance and certification costs are high, as are the costs of transmission of new technology and monitoring of social and environmental standards. Big farm holdings in countries such as Chile are equipped to realize the economies of scale in a way that the small Indian farmer and entrepreneurs are not.

India's Changing Retail and (Agricultural) Marketing Landscape

Vertical integration—the consolidation or co-ordination of production and processing stages by one firm—is a common feature of efficient food marketing systems in most large agricultural economies, but is nascent in India. Only about 4 per cent of output is processed, and only a handful of food processors have annual turnover as large as $150 million—a scale considered small in many other developed and developing countries. Even with more than 1 billion consumers and a retail food market estimated at $133 billion, small 'Mom & Pop' shops still dominate retail food sales. Organized chain stores are emerging and expanding rapidly, but at present account for only about 1 per cent of food sales.

International experience shows that a common organizational pattern in developing countries which have managed to successfully export high-valued agricultural commodities has been the prominence of contract farming and/or vertical integration in the linkages between farm-level production and downstream processing and trade.[21] Both of these were, however, prevented by the Agricultural Produce and Marketing (Development and Regulation) (APM) Act 1951. In 2003, the GoI formulated the model act to reform this act. The model act aims to foster a single market in the country by removing the restriction to sell agricultural commodities wholesale only in state-regulated markets and permitting the private sector to develop and operate wholesale markets. Several state governments have followed suit and amended their APMA to allow for direct marketing, contract farming, and/or developing markets in the private/co-operative sector. While in seven states, the APMA has been amended to allow for all three, in others partial reforms have been undertaken to allow some of these. There are only two states, Bihar and Jharkhand, where there has been no progress on this front.[22] Establishing legal frameworks to protect both farmers and processors in contract farming agreements, and enforcing those agreements is particularly important. With Indian agriculture dominated by small-scale holdings of only about two and a half hectares, food processors struggle to procure adequate supplies of high-quality produce. Contract farming is already expanding in some regions and products, including broilers in Tamil Nadu and Maharashtra and vegetables in Punjab, and has proved successful in reducing marketing risks faced by both buyers and sellers.

Many policies that have, historically, weakened private investment incentives and contributed to India's fragmented, small-scale, and inefficient marketing system are now being changed. The Central Government and several state governments have lifted longstanding measures that restricted private storage and interstate movement of grain and other essential foods. Licences are no longer needed to establish food-processing firms, and regulations restricting their size have been mostly eliminated. In addition, FDI in food processing and marketing—with the exception of retail marketing—is now automatically approved for investments up to 51 per cent equity. The repeal of the Cold Storage Order 1964 in 1998, eliminating the licencing requirement and government control over cold storage fees. To promote storage and warehousing further, the Department of Food & Public Distribution is promoting the development of a negotiable warehouse receipt system in the country to increase liquidity in rural areas. A draft Warehousing (Development and Regulation) Bill, 2005 has been cleared by the cabinet and is expected to be placed before Parliament soon.

There has also been a move to promote the creation of special economic zones (SEZs), some focused on agricultural products. India's Export–Import (EXIM) policy 2003–4 says that corporate sector players with proven credential will be encouraged to sponsor agriculture export zones (AEZs) for boosting agro exports. Appropriate incentives will be provided to them to enable investments in infrastructure, processing, packaging, storage, research and development (R&D), agricultural extension, and other facilities relating to exports in the approved AEZs. As many as 45 AEZs have

BOX 4.4: India's National Horticulture Mission

Recognizing the importance of the horticulture sector in the growth of Indian agriculture as a whole, the Government of India launched an initiative named the National Horticulture Mission to give a boost to this sector. This was launched on 8 July 2004. The main objectives of this mission are: (i) to double horticulture production to 300 million tonnes by 2011–12; (ii) establish convergence and synergy among various ongoing and planned programmes in the field of horticulture development; and (iii) promote the development and dissemination of technologies by blending traditional wisdom and frontier knowledge.

The focus of the Mission will be on horticultural research and development, improving post-harvest management, and promoting processing and marketing. Post-harvest management would include creating suitable infrastructure for efficient post-harvest management and marketing of horticulture produce such as handling, transport, storage, and marketing, besides taking up promotional activities such as dissemination of market information to farmers, processors, traders, and consumers. The government will also provide incentives for setting up horticulture processing industries and food parks in potential areas, besides establishing linkages between markets for the horticulture produce and processing industry. Special thrust would be provided to promote the export of horticultural produce through the establishment of Agricultural Export Zones (AEZs), for which there is a potential global market. The total budget requirement for the Mission during the Tenth Plan period (2002–7) is estimated to be Rs 59.7 billion.

To boost exports, the Ministry of Commerce, has identified a few products, which have the potential to achieve at least 30 per cent annual growth in exports in value terms. These are mango, grape, pineapple, banana, pomegranate among fruits; mushroom, garlic, okra, asparagus, cucurbits, coleenaps, onions among vegetables; rose, chrysanthemum and carnation among flowers. Besides, there are potential customers for other fruits like aonla, guava, sapota, orange, watermelon, copra, cashewnut, nutmeg, cinnamon, clove, cardamom, kokam seeds and tamarind powder. Processed products out of the above also have a demand if processing and packaging follows international standards.

Source: http://www.etagriculture.com/jan_feb2003/policy.html

been notified so far in different parts of the country (also see Box 4.4).

The development of organized retailing is transforming India's food markets by providing an infusion of capital and expertise, as well as promoting linkages and standards backward through the marketing chain. Several large Indian firms have announced ventures in food retailing. FDI has already begun to flow into wholesale food distribution in Bangalore. In order to facilitate the modernization of the retail sector in India, the Foreign Trade Policy 2005–6, has extended concessional duty benefits for capital goods under the Export Promotion Capital Goods (EPCG) Scheme to the retail sector. However, equipment required in the retail supply chain pertaining to material handling, warehousing, and distribution facilities is not included.[23] Another important change has been a partial relaxation on allowing FDI into retail. Earlier, FDI in wholesale cash and carry trading and FDI beyond 51 per cent in export trading were allowed after getting government approval. However, FDI up to 100 per cent is now allowed under the automatic route, that is, without an approval from the Foreign Investment Promotion Board (FIPB), for cash and carry wholesale trading and for single brand products.

To conclude, we find that inefficiencies in supporting services are an important reason for the high protection that is granted to an otherwise competitive sector. The cost advantage that Indian producers (farmers) enjoy when the produce leaves the farm-gate gets slowly eroded during its journey through the supply chain, rendering the product less competitive at the retail end. A large part of the inefficiency in services such as transport, storage, and distribution is due to restrictions on domestic and foreign competition. There are also fundamental weaknesses in the infrastructure in these sectors that can be remedied by creating a favourable environment for private investment and undertaking the necessary public investment. Enhancing the efficiency of supporting service sectors will strengthen both the political case for and economic gains from the liberalization of Indian agriculture trade and improved access to foreign markets.

NOTES

1. While exporters to developed countries (example, grapes to EU) tend to procure directly from farmers, others use many alternative sources of procurement. The exporters surveyed for this study procure 25 per cent of their exports directly from the farmers at the farm-gate and the rest from wholesalers and/or commission agents. The price received by farmers in case of direct procurement is found to be closer to the wholesale level. So while exporters do not seem to enjoy any significant price advantage from direct procurement, the advantage comes in the form of reduced time (a saving of anywhere between 1–3 days) and/or more effective control over the final product.
2. These are indicative numbers and there may be considerable variation across commodities, destination markets, and mode of transport.
3. For example, 50 per cent of US exports today (excluding those to Canada and Mexico) are through air transport.
4. See studies by McKinsey & Co.
5. See Radelet and Sachs (1998).
6. The comparison between Chile and India needs to be qualified. Given its large volume of exports of horticulture products, Chile has a well-established logistics system, dedicated air-freight arrangements, and enjoys the benefit of economies of scale and India has none of those advantages. The comparison however is not so inappropriate if one is looking for a 'global benchmark' that India should strive

to achieve in order to effectively compete in the global horticulture market.

7. The transportation costs for India are based on primary survey data and for others are measured as the difference between the CIF and FOB prices. Although the latter is a widely used measure of international transportation costs (for example, see Radelet and Sachs 1998; Limao and Venables 1999; Clark, Dollar and Micco 2001; UNCTAD 2004), its reliability is often challenged by its critics. There are two reasons why we think our estimates may be a good approximation of the reality. First, the difference between CIF and FOB prices for India is found to be quite similar to the numbers obtained from our primary survey. Second, unlike the previous researchers, we don't use the aggregate CIF and FOB prices to compute the transportation costs; instead we have used commodity-specific and destination market-specific unit values.
8. There are a number of other barriers that impede trade (for example, time to clear customs) and indirectly contribute to higher trade costs. GoI's Task Force on Indirect Taxes had made a number of useful recommendations on custom reforms and those issues well-known to Indian policy makers are, therefore, not covered here.
9. But high transport costs alone cannot explain the sharp difference in market share as the CIF price between Indian and Chilean grapes is not much different. The difference is to some extent explained by quality and standards, which are discussed in Chapter 5.
10. The discussion here has broader implications for the entire international transport sector and is not limited to the transport of horticulture products.
11. Singh and Wallack (2005).
12. Some of these numbers may be out-of-date, and opening up of the sector to private investment and to foreign port operators, such as P&O at JNPT seems to have increased productivity to international levels.
13. Some recent evidence indicates that 70 per cent of services at JNPT are now direct calls, though the same is not true for ports such as Kolkata or Chennai.
14. 88 per cent of container ports in the world conformed to this model of privatization in 1997.
15. Singh and Wallack (2005).
16. This section draws on the World Bank report (2001) 'Forging Subnational Links Transportation and Logistics in South Asia'.
17. FAIDA, CII, and McKinsey & Co (1997); Raghunath and Ashok (2004).
18. Since there is evidence of considerable wastage at the retail-level as well, our primary survey results seem to reinforce McKinsey's finding that the total wastage in the food chain could be as much as 20 per cent.
19. The regulated market holds a key place in India's food chain since all wholesale transactions in agricultural products, by *fiat*, are supposed to take place there. Regulated markets were set up under the Agricultural Produce Markets Act (APMA) to help upgrade market yards, storage, grading and packaging and other market services like market information, intelligence and verification of weights and scales. As on march 2001, there were 7177 regulated market yards in the country.
20. This result is based on a multivariate panel regression where the unit price received by the farmer is regressed against a number of farmer-specific variables including the quantity and quality of infrastructure in their local mandis. The regression also includes state-specific and commodity-specific dummies.
21. Jafee (1992).
22. Information as of 31 January 2006.
23. *Source: FICCI.*

5 Is the External Environment Conducive to Export Expansion?[1]

In this chapter we examine the three broad classes of barriers that affect trade in horticultural products: domestic support, border protection, and differences in standards. We find that the most serious problem is the border protection regime, which is opaque, complex, and deceptively protectionist. The level of direct payments to producers is not significant, while the assistance in the form of price support is another aspect of the protectionist trade regime. Among the most pernicious forms of trade protection in horticulture are the systems of (i) entry prices which discriminate against efficient delivery, (ii) tariff quotas that subjects imports above specified low levels to harsh tariffs, and (iii) SSGs that create considerable uncertainty for successful exporters. Furthermore, preferential access schemes and tariff escalation discriminate against, respectively, imports from certain sources and more advanced stages of processing. Standards present a mixed picture. It is hard to distinguish between the standards that serve legitimate health, safety goals and those that have been captured by protectionist interests. However, some of the most demanding standards today are imposed not by foreign governments but by foreign buyers, reflecting increased consumer sensitivity to quality in all its dimensions.

Given the serious domestic problems identified in Chapter 4, it is difficult to establish how far the external trade regime is today a binding constraint on India's export expansion, but there is little doubt that it would be a serious impediment if India were to emerge as a major exporter. It is, therefore, essential that India takes an aggressive position in the WTO negotiations, seeking not just significantly lower levels of protection, but also much greater transparency, simplicity, and predictability in trade regimes. While India should continue to subject foreign standards to scrutiny, and where necessary, legal challenge, an urgent need in this area is to improve domestic standards and develop mechanisms to signal credibly higher quality.

DOMESTIC SUPPORT AND EXPORT SUBSIDIES

Though information is limited, it seems that governments extend less domestic support to fruits and vegetables than to other agricultural sectors. First of all, direct support to producers is relatively low in Organization for Economic Co-operation and Development (OECD) countries. Of the major industrialized regions, only the EU reports an Aggregate Measure of Support (AMS)[2] for a significant number of fruits and vegetables, while the US, Japan, and Canada indicate moderate levels of AMS support for a few commodities. In the marketing year 2001–2, the EU's direct payments for fruits and vegetables were about Euro 800 million, and took the form of production aid to banana producers and aid to producers of certain products intended for processing (for example, citrus fruits, lemons, grapes, and potatoes). A more significant element of the EU's reported AMS for fruits and vegetables, is the product-specific Equivalent Measurement of Support (EMS)[3] granted through price support or minimum prices. In the marketing year 2001–2, the EMS was estimated to be around Euro 8 billion, with apples and tomatoes the largest beneficiaries—to the extent of around Euro 2 billion each.

In the marketing year 2000, the US made direct payments to producers of apples, onions, potatoes, and tomatoes, but only payments for apples ($174 million) were above the de minimis level. Japan

TABLE 5.1: Domestic Support by European Communities, Marketing Year 2001–2

(Calculation of the Current Total Aggregate Measurement of Support)

Description of basic products	Non-exempt direct payments (including non-product-specific AMS) (Mio Euro)	Product-specific equivalent measurements of support through price support or minimum price (Mio Euro)	Current Total AMS (aggregate) (Mio Euro)
Bananas	212.3		212.3
Apples		2059.50	2059.50
Pears		543.2	543.2
Apricots		120.5	120.5
Cherries		118.2	118.2
Peaches/nectarines		472.3	472.3
Table grapes		216.7	216.7
Plums		69.3	69.3
Lemons		280.7	280.7
Clementines		167.1	167.1
Mandarins		31.8	31.8
Satsumas		16.4	16.4
Oranges		321.0	321.0
Cucumbers		535.1	535.1
Courgettes		170.6	170.6
Artichokes		195.0	195.0
Tomatoes		1944.20	1944.20
Tinned pineapple		6.5	6.5
Citrus fruit for processing	212.8		212.8
Lemons for processing	52.0		52.0
Peaches for processing		62.8	62.8
Plums for processing		41.2	41.2
Pears for processing		30.5	30.5
Figs for processing		6.0	6.0
Tomatoes for processing		366.9	366.9
Grapes for processing	112.7		112.7
Potatoes for processing to starch	212.4	–178.3	34.1

Source: Compiled from WTO notifications.

reported price support for starches ($179 million for potatoes and sweet potatoes in 1998) and direct payments to the vegetable and fruit sectors, but the AMS was below the de minimis level. Similarly, in Canada, only its support for dry beans was above the de minimis level (WTO 2000).

Unlike in many other agricultural sectors, in horticulture the use of export subsidies is not pervasive among the large industrial countries. Table 5.3 shows the most significant export subsidy expenditures based on the notification to the WTO. The amounts are well below those reported for other agricultural categories. In only one country, Switzerland, were expenditures large relative to horticultural exports, with export subsidies accounting for 21 per cent of the country's exports. The EU's export subsidies represented less than 1 per cent of the value of its total exports. Although not notified, the US indirectly subsidized horticultural exports, albeit to a very limited degree, through export credit guarantees. A number of developing countries, including Turkey, Brazil, and South Africa, have also notified export subsidies to the WTO, but these are unlikely to have had a significant impact on world prices.

TABLE 5.2: Domestic Support by US, Marketing Year 2000

(US Product-Specific Aggregate Measurements of Support: Non-exempt Direct Payments)

	Market loss payments	Quality loss payments	($ million)
Apples	169.0	5.4	174.4
Onions	10.0		10.0
Potatoes	11.3	14.4	25.7
Tomatoes	7.2		7.2

Source: Compiled from WTO notifications.

TABLE 5.3: Export Subsidy Expenditures for Horticulture Products

WTO notifying countries	Export expenditures (million $)	Total horticulture exports (million $)	Export exp./ value exports (per cent)
European Union	40	5301	0.8
Switzerland	14	69	20.6
Colombia	13	521	2.4
Turkey	11	2348	0.5

Source: WTO (2000) and subsequent WTO notification updates; Export data from FAOSTAT.

TARIFFS

Many importing countries, in particular the EU, operate complex systems to regulate horticultural imports. In temperate countries, import tariffs and other measures tend to be the lowest and least restrictive on tropical fruits; and highest on temperate fruits and vegetables, particularly during the growing season. The converse tends to be true within many developing countries.

The US, the EU, and Japan use, to varying degrees, similar protection tools: low but highly dispersed ad valorem tariffs, specific duties, seasonal tariffs, entry prices, SSGs, tariff escalation, and preferential access cum tariff-rate quotas. Tariffs for a specific range of products depend on numerous factors, including the date of entry (seasonality factor), the degree of processing (escalation phenomenon), and the relationships with exporting countries (preferential agreements, regional, and bilateral free trade agreements, FTAs).

Average tariffs are low but this is deceptive

Average applied most favoured nation (MFN) tariffs are low in all these countries, ranging from 0.9 per

TABLE 5.4: Applied MFN Tariffs for Fresh Fruits and Vegetables in the Quad Countries

Country		Average and ad valorem rates	All rates average	Percentage of lines covered	Maximum	Standard Deviation	Number of Total Lines
EU	Fresh Fruits	7.3	9.2	75	127.6	15.4	89
(1999)	Fresh Vegetables	5.5	6.8	98	131.8	10.7	200
US	Fresh Fruits	6.1	4.6	100	29.8	7.0	70
(2001)	Fresh Vegetables	4.1	3.1	98	24.3	5.0	189
Japan	Fresh Fruits	8.7	8.7	100	32.0	6.8	56
(2001)	Fresh Vegetables	3.9	3.9	94	40.0	5.6	185
Canada	Fresh Fruits	0.9	0.9	72	8.5	2.5	67
(2001)	Fresh Vegetables	1.1	1.1	74	16.0	2.7	216

Source: WTO IDB Database at the original tariff lone level (6 to 11 digit tariff line depending on the country). Cited in Diop and Jaffee (2004).

cent for fresh fruits in Canada to 9.2 per cent in EU for the same product category (Table 5.4). The average tariffs do not, however, reflect levels of protection for a variety of reasons.

Tariff peaks

Closer examination of the tariff structure highlights the importance of tariff peaks especially in the EU and in Korea. Table 5.5 shows that the US, Canada, and Japan have the lowest tariffs, with, for instance, 85 per cent of the US tariffs under 10 per cent. In sharp contrast, Korea and the EU apply high tariffs on many products. For instance, in Korea, 59 per cent of fresh fruit tariff lines lie between 20 and 50 per cent and 37 per cent of the lines are over 50 per cent. In the EU, 60 per cent of vegetable tariff lines lie between 20 and 50 per cent and 23 per cent of the lines are over 50 per cent. We take a closer look at these peaks for specific products.

TABLE 5.5: Percentage of Tariff Lines at Different Levels/Intervals in Rich Countries: Fresh

Tariff levels	Canada	US	EU	Japan	Korea
Fresh fruits					
Duty free	60	19	2	7	0
1–10%	9	66	49	64	0
11–20%	0	9	22	20	0
21–50%	0	6	0	9	59
Over 50%	0	0	1	0	37
% specific, composite mixed lines	31	1	25	0	5
Fresh vegetables					
Duty free	47	16	0	10	3
1–10%	7	59	11	69	42
11–20%	1	14	1	11	47
21–50%	0	1	60	7	1
Over 50%	0	0	23	0	1
% specific, composite mixed lines	45	9	5	2	5

Note: All figures in percentages. Average applied out-of-quota ad valorem and ad valorem equivalents of non-ad valorem tariffs for those equivalent data reported.
Source: WTO IDB Database.

While the tariff structure is more transparent in most middle-income countries than in the QUAD, the level of average tariff is higher in the former. Average applied MFN tariffs in Morocco, India, and Brazil are far higher than those in the high-income countries analysed above. Table 5.8 presents applied tariff information on some other developing countries—but as we have already noted, average tariffs can be deceptively low. These tariffs pose a potential hindrance to developing country exports,

TABLE 5.6: Percentage of Tariff Lines at Different Levels/Intervals in Rich Countries: Processed

Tariff levels	Canada	US	EU	Japan	Korea
Processed fruits					
Duty free	35	1	0	7	1
1–10%	43	14	0	62	2
11–20%	20	43	0	24	44
21–50%	0	34	47	4	40
Over 50%	0	0	53	4	2
% specific, composite mixed lines	2	7	0	0	10
Processed vegetables					
Duty free	22	3	0	8	3
1–10%	31	39	6	65	6
11–20%	40	50	0	24	76
21–50%	0	6	88	3	5
Over 50%	0	0	3	0	3
% specific, composite mixed lines	7	2	3	0	8

Note: All figures in percentages. Average applied out-of-quota ad valorem and ad valorem equivalents of non-ad valorem tariffs for those equivalent data reported.
Source: WTO IDB Database.

TABLE 5.7: Percentage of Tariff Lines on Fresh Fruits and Vegetables in Selected Developing Countries by Tariff Level

Tariff levels	Brazil	India	Indonesia	Morocco	South Africa
Fresh Fruits					
Duty free	0	0	0	0	15
1–10%	0	0	100	0	47
11–20%	100	3	0	0	29
21–50%	0	94	0	0	9
Over 50%	0	3	0	100	0
% specific, composite	0	0	0	0	10
Fresh Vegetables					
Duty free	24	0	2	0	46
1–10%	0	0	98	0	6
11–20%	76	85	0	13	30
21–50%	0	15	0	19	13
Over 50%	0	0	0	68	0
% specific, composite	0	0	0	0	6

Note: All figures in percentages. Average applied out-of-quota ad valorem equivalents of non-ad valorem tariffs for those equivalent data reported.
Source: WTO IDB Database.

particularly since South–South trade in fruits and vegetables is growing rapidly, and now represents about one-fifth of developing countries' exports.

Entry prices

Simple, uniform tariffs are a rarity in horticulture. For example, under the Uruguay Round Agreement on Agriculture, the EU replaced its reference price system by a system of minimum import or entry prices which apply to fruits and vegetables such as tomatoes, citrus, table grapes, and apples. The entry price system works in conjunction with the system of tariffs (FAO 2004). A band of entry prices is established for each period of importation during the year, climbing in the peak European harvesting period (WTO 2000). The standard ad valorem tariff is charged on imports with a CIF value above the entry price level. A specific duty is applied, in addition to the standard ad valorem tariff, where the import price is below the entry price. This specific duty varies, depending on the level of the import price. To illustrate the complexity, the duty regime for lemons consists of six price bands for the year, the lower end of the price band reaches its highest level from 16 August to 31 October, and the specific component of the duty rises very sharply at prices below the floor in each price band.

Seasonal variation

The seasonal variation in the entry prices itself poses a problem. In effect, the tariffs faced by India's agricultural exports may be much higher

TABLE 5.8: Applied Tariffs in Major Markets of Indian Crops, 2002–3

(simple average in percentage)

		World's Major Importers (Developing Countries)					
HS-6	Product	China	Russia	Indonesia	Malaysia	Mexico	Saudi Arabia
070190	Other potatoes, fresh or chilled	13.0	15.0	5.0	0.0	251.0	
070200	Tomatoes, fresh or chilled	13.0	0.0	5.0	0.0	13.0	
070310	Onions and shallots, fresh or chilled	13.0	15.0	4.4	0.0	13.0	
070810	Peas, fresh or chilled	13.0	15.0	5.0	0.0	13.0	
071290	Dried vegetables, nes	13.0	15.0	4.8	0.0	21.8	5.0
080300	Bananas, including plantains, fresh	13.0	0.0	3.8	0.0	23.0	
080450	Guavas, mangoes and mangosteens, fresh	17.0	5.0	3.0	0.0	23.0	
080530	Lemons and limes, fresh or dried	16.8	0.0	5.0	5.0	23.0	
080610	Fresh grapes	18.4	5.0	5.0	5.0	45.0	
080810	Apples, fresh	14.0	0.0	5.0	5.0	23.0	
200190	Other vegetables, fruits, etc., pres	25.0	0.0	5.0	7.2	23.0	5.0
200950	Tomato juice, unfermented, not contaminated	31.0	0.0	4.2	6.0	23.0	12.0
200960	Grape juice, (incl. must), unferment	23.0	5.0	5.0	4.3	23.0	
200970	Apple juice, unfermented, not contaminated	20.3	10.0	5.0	4.3	23.0	

Source: UNCTAD TRAINS database and WTO IDB CD ROM files.

on account of seasonal tariffs that are designed to protect local producers. To illustrate, the EU tariff on apples ranges from 4.8 per cent + Euro 238/ tonne (April to July) to 11.2 per cent + Euro 238/ tonne (August to December) (Table 5.9 and Figure 5.1). Indian exporters find it difficult to take advantage of the lower tariff as it is applicable during a period of little, if any, production in India. On the other hand, the higher tariff is applicable during the peak harvesting season for apples in India (July to October). There is also substantial seasonal variation in tariffs for tomatoes in the EC and US, and for grapes in the US and Japan. In Japan, ad valorem tariffs for fresh grapes rise from 7.8 per cent between November and February to 17 per cent between March and October—the peak harvesting season in India being between February and July.

Table 5.9 makes an inevitably inadequate attempt to convey the scale and variation in protection. For example, tomatoes receive an average protection in the EU estimated at the 6 digit level to be only 7.4 per cent. But if Indian imports arrive at the wrong time and at a low unit value, they could end up paying an ad valorem equivalent of as much as 100 per cent.

Tariff rate quotas

Next, we consider the impact of tariff rate quotas (TRQs) which were established in the Uruguay Round. Lower tariffs are applied to in-quota

TABLE 5.9: Seasonal Variation: Current MFN Import Tariffs for Selected Fruits and Vegetables—EC, US, and Japan

MFN applied tariffs	Japan		EC		US	
Apples						
Fresh	17		11.2	(Aug 1–Dec 31)	0	
			+ 238€/t			
			6.4	(Jan 1–Mar 31)		
			+ 238€/t		0.74	
Dried	9		6.4		c/kg	
Grapes						
Fresh	17.0	(Mar 1– Oct 31)	11.5	(Dec 1–July 31)		(Feb 15–Mar 31)
			14.5	(Aug 1–Oct 31)	Free	(Apr 1–June 30)
	7.8	(Nov 1–end Feb)				
			+ 96€/t			
			11.5	(Nov 1–Nov 30)		
			+ 96€/t			
Dried	1.2		2.4		3.5	
Tomatoes						
Fresh or	3		8.8	(Nov 1–May 14)	3.9	(Mar 1–Jul 15)
chilled			+ 298€/t		3.9	(Sep 1–No 14)
					2.8	(Jul 15–Aug 31)
			14.4	(May 15–Oct 31)	2.8	(Nov 15–end Feb)
			+ 298€/t			

Note: Value for customs purposes.
Source: FAO (2004) and staff compilations.

shipments, and much higher tariffs to out of quota shipments. Fruits and vegetables turn out to be particularly infested by TRQs, with 355 TRQs in place, which is a quarter of all such measures (Table 5.1). Norway has as many as 116 TRQs, followed by Poland and Hungary with 37 and 33, respectively. From the Indian perspective, the 25 measures maintained by the EU are of the greatest significance, for which Table 5.16 presents detailed information. In addition to tariff quotas registered at the time of the Uruguay Round, the EU also establishes country-specific, generally duty-free, preferential quotas for certain products, such as potatoes, dried onions, and sweet peppers from which India does not benefit. In respect of most products, the EU does not have country-specific allocations, except in bananas, where historical shares are a determinant of individual country quotas.

The available evidence, and discussions with Agricultural and Processed Food Products Export Deveopment Authority (APEDA) officials indicate that the current level of exports of fruits and vegetables from India are in most cases below the quota limit and hence many of the tariff restrictions imposed by the EU are non-binding on India. For example, with regard to bananas, EU's tariff quota limit is 2.2 million metric tonnes, while India's exports to the EU in 2002 were 20 metric tonnes, worth $73,000, while a country like Ecuador exported bananas worth $683,204,000. Similarly, the EU's non-specific quotas on potatoes and apples were, respectively, 4000 and 1500 metric tonnes, whereas India's exports were 1 and 2 metric tonnes, respectively. One area where Indian exports exceed the level of the tariff quota is for table grapes: the quota is 10,000 metric tonnes whereas India's exports in 2002 were 16,491 metric tonnes.

Special safeguards

The possibility of resort by an importing country to SSGs, can undermine market access. Under the

Figure 5.1a: Entry Price for Lemons and Tomatoes into the EU, 2004–5

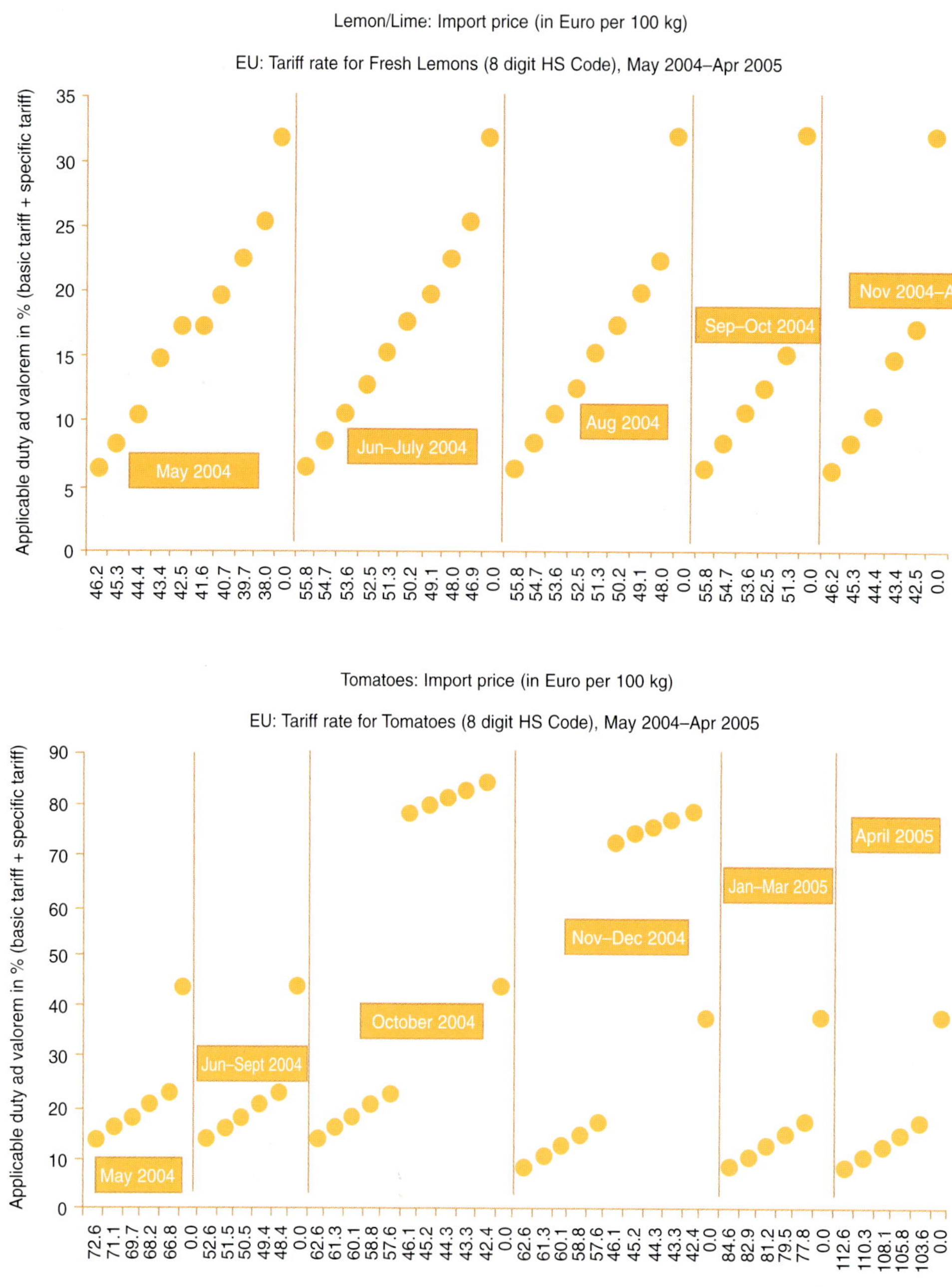

Figure 5.1b: Entry Price for Apples and Grapes into the EU, 2004–5

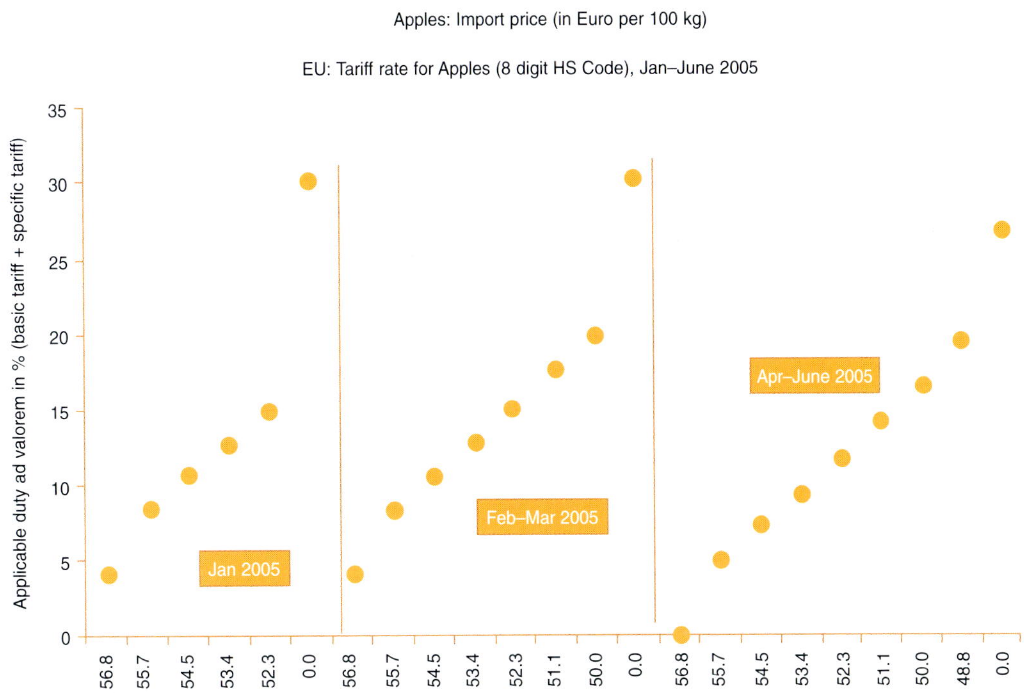

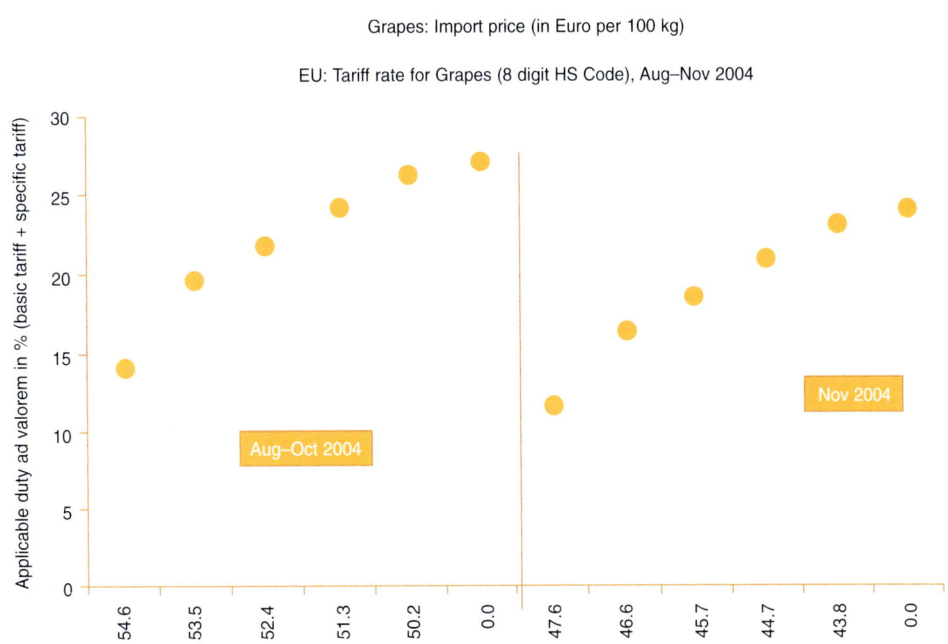

Source: WTO IDB Database.

Uruguay Round Agreement on Agriculture (URAA), a number of countries have reserved the right to implement SSGs if the import price of certain products falls below a 'trigger price', or if the quantities being imported exceed a 'trigger volume'. SSGs have been registered for all those products to which entry prices apply. The relationship between trigger prices and entry prices varies depending on the commodity and the time of the year. In practice, many trigger prices are well above entry prices, limiting the relevance of the SSGs to importers. But a number of products in the fruits and vegetables category have been notified for taking SSGs, especially volume-based SSGs. For example in their latest notification to the WTO, the EU specifies that it too had volume-based SSGs in the period 2001–2 on a wide range of products, including tomatoes, cucumbers, artichokes, courgettes, oranges, clementines, mandarins, tangerines, satsumas, wilkings and similar citrus hybrids, lemons, table grapes, apples, pears, apricots, cherries, peaches, including nectarines and plums. Price-based SSGs are rarer for fruits and vegetables, for example such action has been taken by Japan for peas and Poland for tomatoes.

TABLE 5.10: EU and US Applied and Specific Tariffs on Selected Indian Fruits and Vegetable, 2004

(in per cent)

HS-6	Product Name	Applied Tariffs at HS-6	No. of HS-8 Lines	Tariff Rate at HS-8 Minimum	Tariff Rate at HS-8 Maximum
EU Markets:					
070190	Potatoes, fresh or chilled	10.1	10	5.8	13.4
070200	Tomatoes, fresh or chilled	7.4	2	8.8	14.4% + 298 EUR/mt
070310	Onions, fresh or chilled	9.6	6	9.6	9.6
070810	Peas, fresh or chilled	9.9	8	8.0	13.6
080300	Bananas, including plantains, fresh	16.0	6	680 E/1000 kg/net	680 E/1000 kg/net
080450	Guavas, mangoes, and mangosteens	0	2	0	0
080530	Lemons and limes, fresh or dried	12.8	4	6.4	6.4% + 256 EUR/mt
080610	Fresh grapes (seedless)	15.5	10	11.5	17.6 + 96 EUR/mt
080810	Apples, fresh	0	8	0	6.4% +238 EUR/mt
US Markets:					
070190	Potatoes, fresh or chilled	0	30	0.3 c/kg	1.7 c/kg
070200	Tomatoes, fresh or chilled	0	45	22.4 c/kg	6.6 c/kg
070310	Onions, fresh or chilled	0	45	0.8 c/kg	5.5 c/kg
070810	Peas, fresh or chilled	0	31	0.5 c/kg	8.6 c/kg
080300	Bananas, including plantains, fresh	0.5	45	0	1.4
080450	Guavas, mangoes, and mangosteens	0	45	1.5 c/kg	33.1 c/kg
080530	Lemons and limes, fresh or dried	0.8	46	\$0.02/kg	\$0.025/kg
080610	Fresh grapes	0	46	\$1.13/m^3	\$8.83/m^3
080810	Apples, fresh	0	15	1.1 c/kg	1.1 c/kg

Source: WTO IDB database.

TABLE 5.11 Number of Tariff Quotas by Member and Product Group

Member	CE Cereals	OI Oilseeds	SG Sugar	DA Dairy	ME Meat	EG Eggs	BV Beverages	FV Fruit & Vegetables	TB Tobacco	FI Fibres	CO Coffee	O Other	All
Australia	–	–	–	1	–	–	–	–	1	–	–	–	2
Barbados	1	1	1	2	4	1	1	25	–	–	–	–	36
Brazil	1	–	–	–	–	–	–	1	–	–	–	–	2
Bulgaria	6	2	4	6	26	–	5	15	2	–	7	–	73
Canada	4	1	–	11	4	1	–	–	–	–	–	–	21
Chile	–	–	1	–	–	–	–	–	–	–	–	–	1
China	4	3	1	–	–	–	–	–	–	2	–	–	10
Chinese Taipei	1	1	1	1	5	–	–	13	–	–	–	–	22
Colombia	13	20	4	5	17	–	–	4	–	2	–	2	67
Costa Rica	3	–	1	13	6	1	–	1	1	1	–	–	27
Croatia	1	–	3	–	3	–	–	–	–	–	2	–	9
Czech Rep.	2	5	1	4	5	–	4	2	–	–	–	1	24
Dominican Rep.	2	–	1	1	1	–	–	3	–	–	–	–	8
Ecuador	6	2	3	1	2	–	–	–	–	–	–	–	14
El Salvador	1	–	1	4	4	–	–	–	1	–	–	–	11
EC-15	15	–	3	12	28	3	1	25	–	–	–	–	87
Guatemala	6	7	1	1	3	–	–	3	1	–	–	–	22
Hungary	7	4	2	4	8	1	4	33	1	–	6	–	70
Iceland	17	22	3	4	13	1	1	18	–	–	5	6	90
Indonesia	1	–	–	1	–	–	–	–	–	–	–	–	2
Isreal	1	–	–	4	3	–	–	4	–	–	–	–	12
Japan	4	1	–	12	–	–	–	1	–	1	–	1	20
Korea	15	5	2	5	7	1	1	20	–	2	4	5	67
Latvia	3	–	1	–	–	–	–	–	–	–	–	–	4
Lithuania	–	1	1	–	2	–	–	–	–	–	–	–	4
Malaysia	1	–	1	2	10	2	–	1	1	–	1	–	19
Mexico	3	–	1	2	2	–	–	2	–	–	1	–	11
Morocco	5	6	1	1	3	–	–	–	–	–	–	–	16
New Zealand	–	–	–	–	–	–	–	2	–	–	–	1	3
Nicaragua	3	1	1	1	2	–	–	1	–	–	–	–	9
Norway	37	2	2	14	32	3	1	116	–	–	3	22	232
Panama	2	–	–	11	2	–	–	4	–	–	–	–	19
Philippines	2	–	1	–	9	–	–	1	–	–	1	–	14
Poland	12	4	2	8	14	3	5	37	1	3	11	9	109
Romania	1	–	–	3	1	–	4	1	1	–	1	–	12
Slovak Rep.	2	5	1	4	5	–	4	2	–	–	–	1	24
Slovenia	4	1	–	2	12	–	–	1	–	–	–	–	20
South Africa	11	8	3	6	5	1	1	12	1	1	4	–	53
Switzerland	3	–	–	2	6	3	3	9	–	–	–	2	28
Thiland	2	6	1	2	–	–	–	5	1	1	4	1	23
Tunisia	3	–	1	3	3	–	–	3	–	–	–	–	13
United States	1	2	6	24	1	–	–	5	1	7	7	–	54
Venezuela	20	19	3	6	10	–	–	–	–	–	1	2	61
All Members	217	124	51	181	247	21	35	355	13	18	56	53	1425

Source: OECD Agricultural Market Accers (AMAD) database.

TABLE 5.12: TRQs on Fruit and Vegetable Imports into the EU[1,2]

Fruit/Vegetable	TRQ
Bananas[3]	3,403,000
Oranges	20,000
Minneolas	15,000
Lemons	10,000
Table grapes	1500
Applies	600
Apricots	2500
Cherries	800
Pears	1000
Potatoes	4000
Carrots	1200
Sweet peppers	500
Mushrooms[4]	62,660
Onions, dried	12,000
Almonds, excl.	45,000
Orange juice, frozen	1500

Notes: 1. Notified under the URAA as current access quotas and non-tariffied product quotas. NB. No minimum access quotas were negotiated for fruits or vegetables.
2. In addition to the scheduled tariff quotas, the EU also establishes additional preferential quotas. For example, in 2004, these included an additional 5880 tonnes for dried onions (0 per cent duty vs 10 per cent in quota duty); 6200 tonnes for carrots (0 per cent and 2.7 per cent vs 7 per cent) for Switzerland and Slovenia; 9500 tonnes for potatoes (0 per cent and 1.9 per cent vs 3 per cent) for Bulgaria, Lebanon and Slovakia.
3. Includes the 2.2 million metric tonnes tariff quota established under the URAA and a further preferential quota of 1.103 tonnes.
4. Mushrooms, prepared or preserved. Tariff level depends on mushroom type.
Source: FAO (2004).

Preferential regimes

India is disadvantaged in foreign markets when exporters from other countries are exempted from tariffs that Indian exporters must pay. The problem is most visible in the EU, which has concluded preferential trade agreements with important horticultural producers, quite apart from the unilateral preferences that it has granted to some of the poorest countries. Thus Turkey enjoys duty-free access in a range of products where India must pay average tariffs of between 5 and 10 per cent—but note, from our discussion above, that the actual wedge might be much greater. A similar problem exists in the US market vis-à-vis Mexico, but it is not clear how far India would be an effective competitor even in the absence of the tariff advantage that Mexico enjoys.

Tariff Escalation

The tariff structure in the EU, the US, and other countries (for example, Japan and Korea) also features a high degree of escalation, which has the effect of creating a disincentive to the export of relatively more processed products. The average applied tariffs in Table 5.15 do not convey the full impact of escalation in specific products, but note that tariffs on processed products are consistently higher than those on semi-processed products, and in the EU (but not the US) the tariffs on semi-processed products are higher than those on agricultural raw products. Tariff escalation is particularly pronounced in fruits in the EU, where only about 1 per cent of fresh fruit tariff lines are subject to duties above 20 per cent, but 100 per cent of the processed fruit tariff lines lie above 20 per cent. The low US tariffs do not exclude the use of high levels of protection for particular products, as the US protection of its own orange juice industry illustrates.

Summing up

The combination of seasonal variation, entry prices, TRQs, specific tariffs, tariff peaks buried

TABLE 5.13: EU's Applied Tariffs on Selected Crops by Selected Country, 2003

(simple average in per cent)

HS-6	Product Name	World	India	Morocco	South Africa	Turkey
070190	Potatoes, fresh or chilled	6.23		7.1	0.8	3.29
070200	Tomatoes, fresh or chilled	7.36		4.32	10.9	0
070310	Onions, fresh or chilled	5.65	6.1	3.05	7.3	4.8
070810	Peas, fresh or chilled	3.03	6.37	6.37	6.37	0
071290	Dried vegetables, nes	4.84	7.61	0.99	7.35	0
080300	Bananas, including plantains, fresh	6.36	12.5			0
080450	Guavas, mangoes and mangosteens	0	0		0	0
080610	Fresh grapes	8.2	9.43	2.79	9.05	5.15
080810	Apples, fresh	2.49			2.9	0
151620	Vegetable fats and oils	3.96	4.8	4.8	0	0
200190	Other vegetables or fruits	6.81	10.46	5.21	6.37	0
200950	Tomato juice	10.99				0

Source: UNCTAD TRAINS Database.

TABLE 5.14: US Applied Tariffs on Selected Crops by Selected Country, 2003

(simple average in per cent)

HS-6	Product	World	India	Chile	Mexico
070190	Potatoes, fresh or chilled	0	0		
070200	Tomatoes, fresh or chilled	0		0	0
070310	Onions, fresh or chilled	0	0	0	0
070810	Peas, fresh or chilled	0		0	0
071290	Dried vegetables, nes	4.4	3.85	3.85	1.05
080300	Bananas, inluding plantains, fresh	0.04	0		0
080450	Guavas, mangoes, and mangosteens	0	0		0
080610	Fresh grapes	0		0	0
080810	Apples, fresh	0		0	
151620	Vegetable fats and oils	6.82	7.7	7.7	0
200190	Other vegetables or fruits	4.28	2.32	2.32	0
200950	Tomato juice	0		0	

Source: UNCTAD TRAINS Database.

TABLE 5.15: Tariff Escalation of Agricultural Products in US and EU, 2003

(simple average in per cent)

Stage of processing	United States Applied Rate	United States GSP Rate	United States FTA Rate	EU Applied Rate
Agricultural raw products	6.9	4.5	0.5	3.5
Agri. semi-processed products	4.4	1.1	0.3	7.0
Agricultural processed products	9.9	6.6	0.7	10.9

Source: WTO IDB CD ROM 2004.

at the 8 or higher digit level of disaggregation, and SSGs creates a nightmarish trade regime. In particular, the entry price regimes (which operate for tomatoes, lemons, grapes, and apples) create a strong disincentive to export at a low price because the specific tariff is designed to increase the price of imports to the entry price level. Thus, the import regime itself is designed to dampen the incentive for efficient delivery.

STANDARDS

Differences between domestic and foreign standards have been identified as an important barrier to trade in horticulture but the evidence of their impact on India's exports is mixed. SPS measures are not a major barrier for the bulk of fresh exports that are directed towards regional markets with relatively low standards. But some of the attempts to penetrate higher income markets have been impeded by the failure to comply with required standards. Our survey-based approach can only identify some of the issues that arise in a complex area like standards. Our findings are nevertheless consistent with a more detailed World Bank study on the role of standards in India's horticultural trade which finds that the challenges posed by standards have manifested themselves in different ways for Indian horticulture, including: prohibitive barriers to accessing particular markets; temporary losses due to the rejection of consignments that do not meet standards; higher consignment-specific or recurrent transactions costs due to duplicative testing, inspections or further treatment; and patterns of defensive commercialization whereby firms fail to pursue potentially profitable opportunities because they believe that they are incapable of complying with the relevant standards.

A significant constraint on horticultural trade in general are the SPS controls imposed by destination buyers. These are particularly stringent in the US and Japan. Between 1995 and 2000, it is estimated that nearly 270 SPS measures were introduced on imports of fruits and vegetables. The most prominent case of prohibitive barriers to accessing particular markets involves fresh mango exports to Australia, Japan, and the United States. Quite apart from the high level of standards, the lack of harmonized technical standards and treatments may be a hindrance to trade (FAO 2004). For example, some countries apply the Codex Alimentarius standards for maximum (pesticide) residue levels (MRLs), while other countries apply their own, often stricter MRLs which may only partially conform to Codex.

In the EU, marketing standards are in operation for a wide range of fresh horticultural produce (FAO 2004). The standards apply at all stages of distribution, and include imports and exports. The standards apply to products produced within the EU, so tropical fruits and vegetables, for example, are excluded. Produce exported from a non-EU origin must meet the standards applied to domestic produce, in respect of quality, packaging, and labelling. Products that meet the required marketing standards are issued with a conformity certificate; those failing to meet the standards are refused entry. Some countries are able to carry out their own conformity checks. Only two of these, Morocco and South Africa, are non-European.

Japan's phytosanitary barriers are a far more important determinant of what can and what cannot enter the market than its import tariff regime (FAO 2004). All consignments of fresh fruit and vegetables require a phytosanitary certificate and are subject to import inspection on arrival. Dried fruits are subject to import inspection but are excluded from plant quarantine. Imports of a number of fresh vegetables are banned from most countries because of plant disease restrictions. Fruits such as mangoes, papayas, and lychees may be imported from specified origins (principally China, the Philippines, and Thailand) provided they receive vapour heat and/or cold treatments.

Other products are affected by the policy of fumigating any fresh produce shipments where insects are seen to be present, even if these are already endemic in Japan. Fumigation can seriously damage the quality of imported produce, and this has been a problem in the case of light coloured vegetables such as lettuce and cauliflowers. In the case of the organic sector, random fumigation of organic fresh produce presents a major disincentive to export organic produce to Japan, because once fumigated the product can no longer be sold as organic. Exporters have experienced problems even where hot air treatments have been applied and the products have been cleared by quarantine arrangements in the country of origin.

In the United States, all products are subject to examination by the Food and Drug Administration (FDA) when they enter (FAO 2004). By law, all products must meet the same standards as domestically produced products. The Animal and Plant Health Inspection Service (APHIS) inspects imported agricultural products. Importers must obtain phytosanitary certificates from the country of origin. Quarantine laws require a cargo to be inspected at the first US port of arrival. Where pests are discovered, unless they are endemic, APHIS will undertake fumigation. Where APHIS determines that imports of a certain product (often from a specific origin) may lead to the introduction of new, or not widely distributed, plant pests into the United States it introduces a ban.

Somewhat surprisingly, our survey of Indian exporters reveals contrasting experiences with standards. As Figure 5.2 reveals, less than half reported a decline in exports as a consequence of rising standards. Similarly, only a few reported making adjustments in the product, its packaging, and even shifting sales away from certain markets. A large number of exporters did not however report any response at all to higher standards. One explanation is that the bulk of India's exports are to regional markets where standards are lower (reflecting the pattern of defensive specialization noted already), and the small proportion that was directed to industrial country markets did in fact suffer adverse consequences because of demanding standards. There are two other reasons, however, why our survey may not tell the whole story. First, there is a selection bias in that we were only able to interview exporters regarding conditions in markets where they were 'able to export'. Any exporters who may have been obliged to exit as a consequence of prohibitively costly standards were not part of the sample. Second, our survey focused on the impact of changing standards rather than already high standards, which could deter exporters without any change in their level.

There is other evidence to suggest that standards do pose a problem in industrial country markets. In the WTO's SPS Committee, India has supported several complaints raised against measures

Figure 5.2: Impact of, and Response to, Higher Standards in Destination Markets, 2004–5

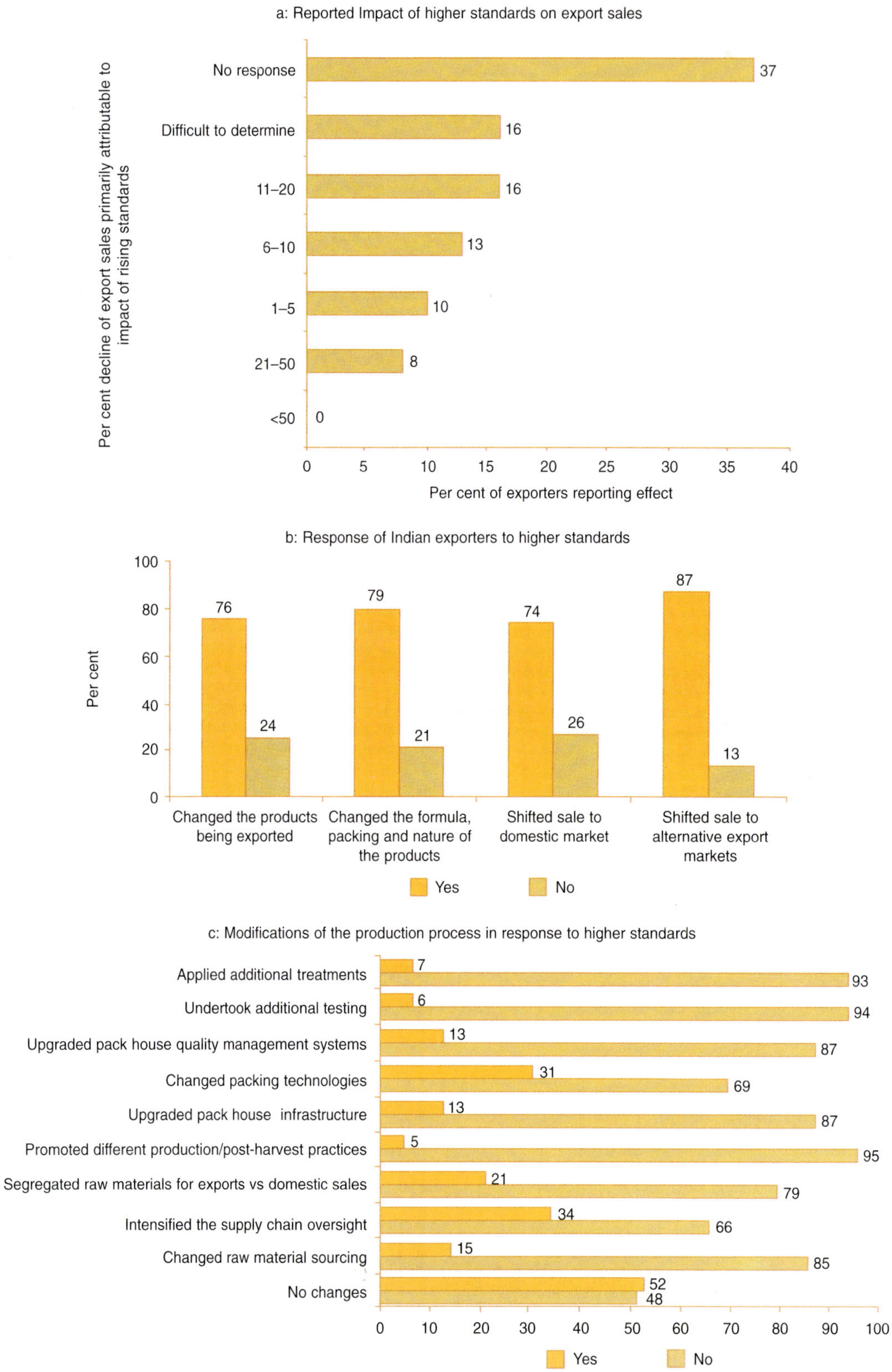

Source: Survey of exporters (2004–5).

regulating imports of fresh and processed fruits and vegetables (Roberts and Krissoff 2004). These include complaints against the EC on sampling methods for aflatoxins in dried fruits and other products; against Australia on import restrictions on durian and tropical fresh fruit (such as mangoes, bananas, and pineapples). Furthermore, in the year 2003, Indian grapes were found to be in violation of EU pesticide residue tolerances. The government responded by imposing a set of mandatory controls and product checks designed to exclude products that did not meet EU norms, which appear to have been successful because no further adverse tests have been reported in the last two years. However there remain outstanding technical and institutional issues that need to be resolved, and in the longer term, the private sector will need to address the standards-related challenges facing the industry.

A case study of the experience of ten mango exporting firms reveals that they face at least seven different types of non-tariff measures (NTMs) in overseas markets (Wignaraja et al. 2001).[4] These include packaging and labelling regulations, pesticide residue limit regulations, chemical content limitations, fruit fly related regulations, uniformity requirements, labour standards, and documentation procedures. Each of these regulations has a negative effect on Indian exporter's performance in their respective export market. The imposition of NTMs follows a specific pattern. Industrial countries such as the US, EU, and Japan impose sophisticated regulations like registration, packaging and labelling, pesticide residue limits and aflatoxin content, fruit fly, and labour standards. On the other hand, the problems faced by exporters in the Gulf countries are caused mostly by uniformity in size of product and documentation procedures (Table 5.17 lists the barriers).

It is also important to note that statutory food safety policy instruments are being overtaken by a large number of private standards imposed by global retailers and processors, led by supermarkets. Jaffee (2003) draws an important distinction between compliance standards which are legally mandated and those which are a commercial requirement. Supermarket standards are usually even higher.[5] All export firms must now have sophisticated quality assurance systems in place to document seed procurement, planting schedules, agrochemical and fertilizer use, and which ensure full traceability through the supply chain (FAO 2004).

The importance of buyer standards is revealed in Figure 5.3a, which shows that most of the problems have arisen not in the form of restrictions by foreign authorities, but warnings and reductions in price and demand from foreign buyers. Similarly, Figure 5.3b reveals that by far the most important factor was demands from external buyers.

Compliance with standards can be costly. In our sample, one in ten firms reported a cost increase greater than 11 per cent, and two-thirds reported at least some cost increase (Figure 5.4). Table 5.8 shows how the mango firms surveyed in Wignaraja et al. (2001) have responded to the standards. The estimates of the costs of compliance are patchy, but do confirm that in some cases there are likely to be increases in fixed costs, variable costs, or both. That the cost of meeting standards is potentially high is also consistent with the findings of Jaffee (2003) that one large Kenyan producer/exporter expects to spend around $300,000 per annum, or 3 per cent of turnover, on annual food safety management costs. Collinson (2001), writing on the Kenyan flower industry, reports that the management time to implement compliance, in the form of documentation and record keeping, is

Figure 5.3: Signals that Reflect a Failure to Meet Rising Standards and Factors Stimulating a Positive Response, 2004–5

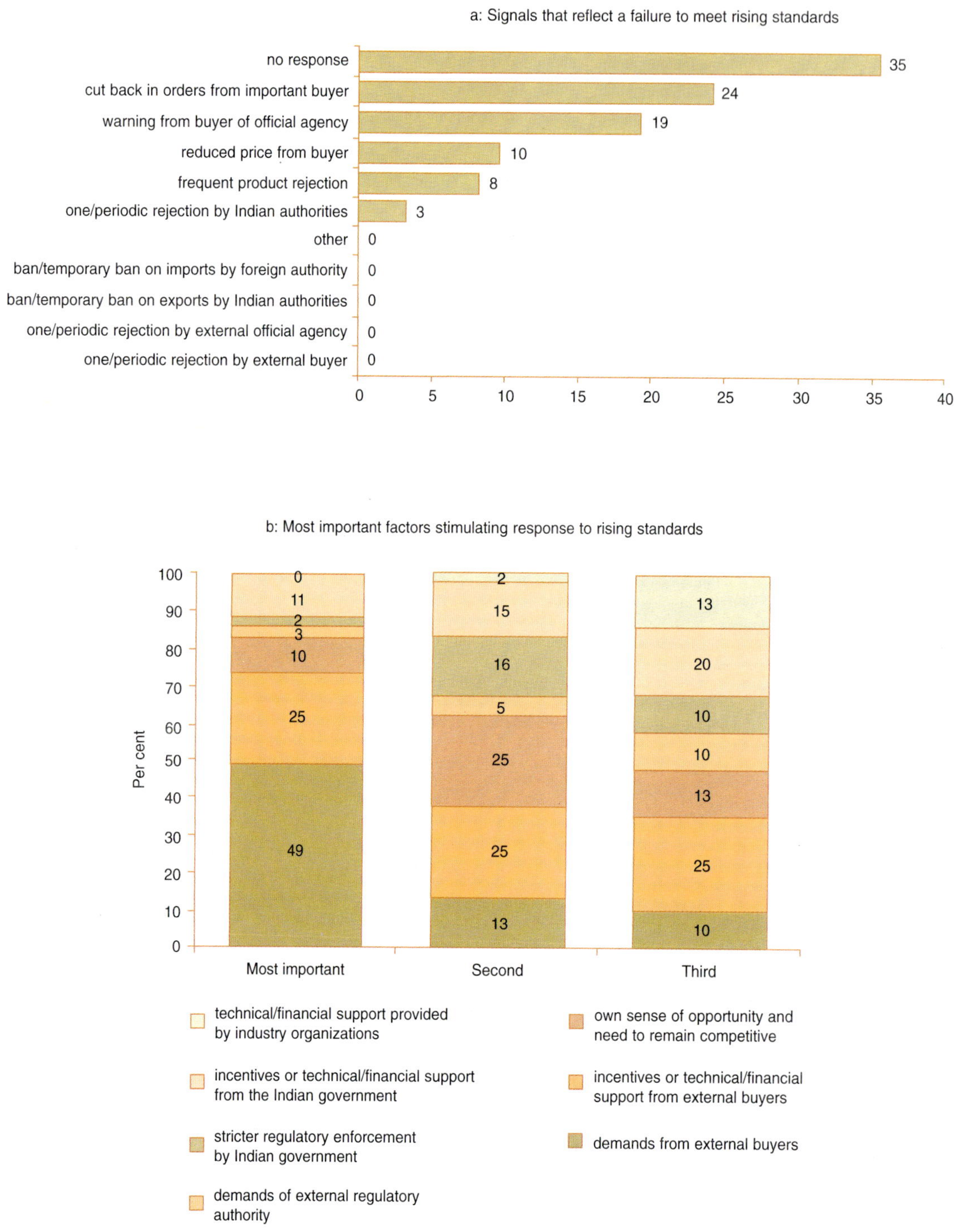

Source: Survey of exporters (2004–5).

BOX 5.1: India's Domestic Standards on Agricultural Products

In India, till the mid-1990s, the agricultural sector was fairly closed to imports. As such, few standards were applied to imports. The main regulations were meant primarily to ensure that imports were subject to the same quality standards governing Indian products. In 2003, however, the continual rejection of its food consignments by the EU and the US on grounds of SPS measures brought these issues into sharp focus. The problems of lack of awareness, and of the consistency of domestic SPS standards with the standards of the importing countries came up in a major way. Apart from prompting India to raise domestic standards, especially for export items, it also spurred it to begin putting in place stringent standards of its own. This was formally done with the enactment of the Plant Quarantine (Regulation of Import into India) Order 2003.

At present, the important regulations governing import of agricultural and food products into India are as follows:

- Import of all food products, domestic sale and consumption of which are governed by the prevention of Food Adulteration Act, 1954, are subject to all the conditions laid out therein.
- Since 2000, import of products like food preservatives and additives is subject to compliance of mandatory Indian quality standards as applicable to domestic goods. For compliance of these requirements, all manufacturers/exporters of these products to India are required to register themselves with the Bureau of Indian Standards (BIS).
- Pre-packaged products have to meet strict labelling requirements. For imported products, the name and address of the importer must be indicated on the label, as also the quantity in standard units of weights and measures, month and year of packing, and the maximum retail sale price at which the commodity will be sold to the ultimate consumer.
- Import of all edible/food products is also subject to the condition that, at the time of importation, the products have a valid shelf life of not less than 60 per cent of its original shelf life.
- Import of some commodities requires an agricultural permit. These commodities include cereals and pulses, and all other agricultural commodities meant for consumption. This permit has to be obtained prior to import.
- Import of all primary agricultural products is subject to a Bio-security and SPS Import permit to be issued by the Department of Agriculture. According to regulations, this permit will be based on import risk analysis of the product, to be conducted on scientific principles, in accordance with WTO agreements.
- Every exporting country has to provide a certificate saying that the exported product has been devitalized, and this has to be endorsed in the phytosanitary certificate.

In principle, these standards seem reasonable, and there is little doubt that better SPS standards should lead to the reduction of health risks and should benefit consumers. However, some exporters have complained that these SPS rules, technical standards, and registration rules are sometimes applied in a non-transparent manner which hurts imports. Another complaint has been that food safety laws need streamlining and aligning with international standards. Indian food law now falls under five outdated statutes, with jurisdiction spread across four ministries, thus greatly increasing the cost and complexity of compliance.

Source: BIG's Easy Reference Customs Tariffs 2005–6, Arun Goyal.

Figure 5.4: Impact of Rising Standards on Production Costs, 2004–5

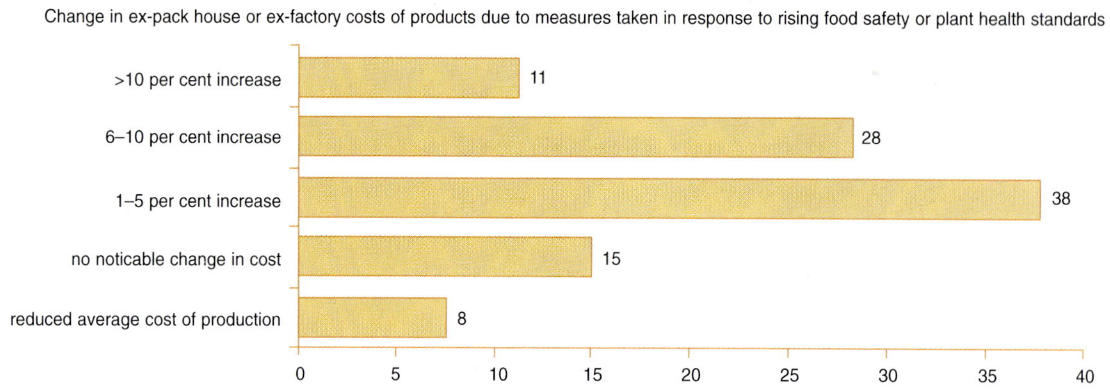

Source: Survey of exporters (2004–5).

one of the biggest costs faced by growers and falls disproportionately on small growers.

It would be wrong to describe all standard-related problems as a manifestation of protectionism in industrial countries. Inadequacies in domestic standard setting legitimize foreign barriers. Indian food law now falls under five outdated statutes, with jurisdiction spread across four ministries, thus greatly increasing the cost and complexity of compliance. Concerns have also been expressed about weaknesses in the enforcement of pesticide-related regulations and ensuring standards of hygiene, which undermine the credibility of domestic certification processes. India today has a plethora of regulatory authorities tackling these issues, often in a piecemeal manner. They include various central ministries (Commerce, Food and Consumer Affairs, and Food Processing), export promotion institutions, standard-setting bodies (24 of them), and a host of regulating bodies at the state level. Other important authorities include the BIS, National Accreditation Board for Testing and Calibration Laboratories (NABL), National Quality Campaign (NQC), Central Committee of Food Standards (CCFS), and Standardization Testing and Quality Certification (STQC). What is needed perhaps is integration among these bodies, harmonization, and a national notification of standards. A major revamp of the food law aimed at consolidation of responsibilities and jurisdiction, as well as closer links to international standards, is now underway, although there is no clear time frame for its completion.

In fact, standards need to be seen not only as an impediment but also as an opportunity, as has been argued by Jaffee and Henson (2004). They show that rising standards accentuate underlying supply chain strengths and weaknesses, and thus impact differently on the competitive position of individual countries and different market participants. Some countries and industries are even using high quality and safety standards to successfully position themselves in global markets. The challenge for countries like India is to take a proactive rather than a defensive approach, so that they can exploit their strengths and overcome their weaknesses to benefit from the emerging commercial and regulatory context. Some evidence for this more optimistic view is to be found in our survey of exporters (Figure 5.5). Fifteen per cent of exporters reported that rising standards had a beneficial impact on their exports, which was only slightly lower than the 18 per cent who reported an adverse impact.

Finally, it may be of some interest to compare exporters' evaluation of the different types of barriers they face in the four markets where most of them operate: Bangladesh, East Asia, Europe, and the Middle East. NTBs emerge as more important than customs duties in all markets, and are comparable in significance with international transport costs. It is also evident that payment problems are a serious issue in all markets but are most acute in Bangladesh,

Figure 5.5: Overall Impact of Rising Standards on Exports, 2004–5

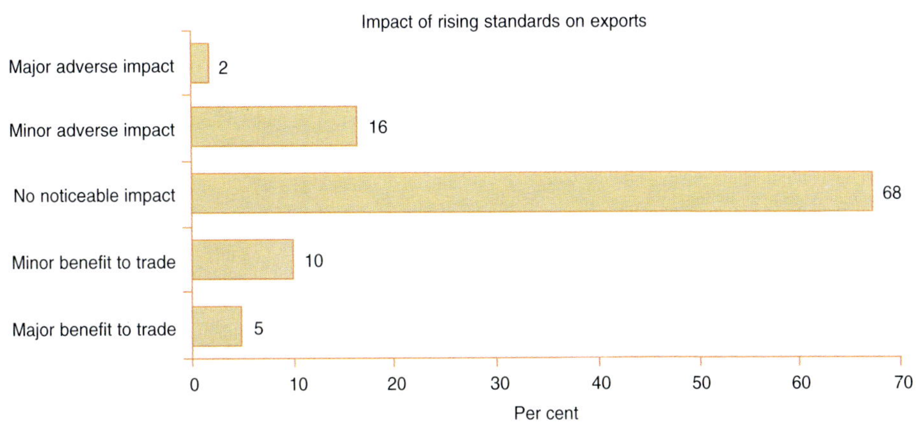

Source: Survey of exporters (2004–5).

Figure 5.6: Evaluation of Barriers in Four Destination Markets, 2004–5

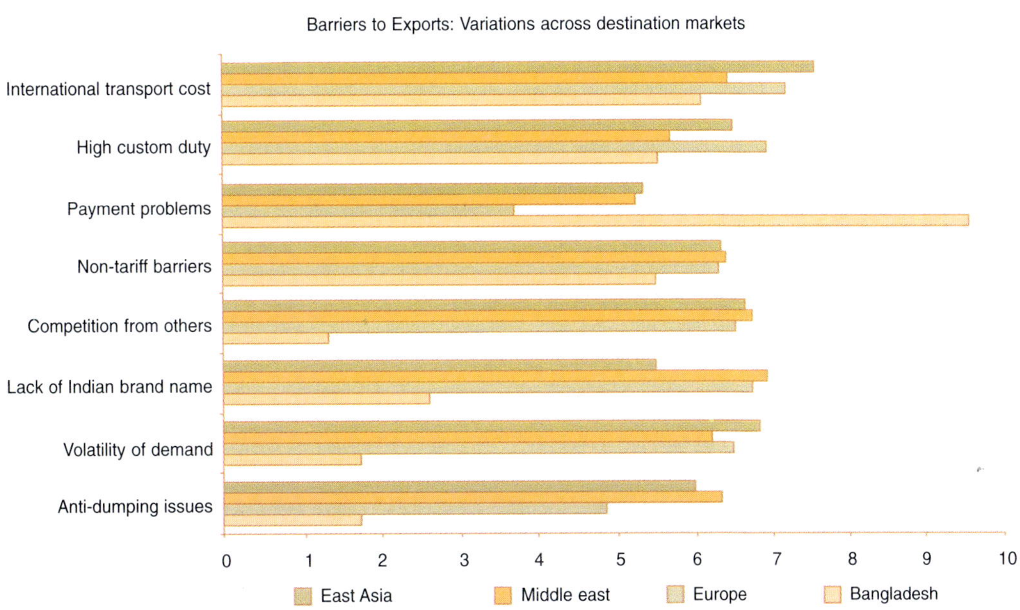

Source: Survey of exporters (2004–5).

TABLE 5.16: Detailed Information on the EU's Tariff Rate Quotas

1	2	3	4	5	6	7	8	9	10	11	12
							Bound Tariffs		Total Imports		
Description	REG	HSNo	Units	Quota	Specific (1) × Global (0)	Trade Weighted applied tariffs (per cent)	In-Quota (TI) (per cent)	Out-of Quota (T2) (per cent)	Value	Quantity	Fill Rate Truncate (per cent)
Manioc (casava)	LD	071410	TNE	5,500,000	1	20	6	84	201,350,760	2,922,368	26
Manioc (casava), other than pellets...	LD	ex071410,ex071490	TNE	1,352,590	1	3981	6	49	4,504,696	14,392	1
Sweet potatoes (not China)	LD	07142090	TNE	5000	1		0	16			0
Sweet potatoes (China)	LD	07142090	TNE	600,000	1	7	0	16	194,886	164	0
Fresh bananas, other than plantains	HI	08030019	TNE	2,200,000	1	147	19	172	1,855,011,345	3,283,588	100
Oranges	FC	ex080510	TNE	20,000	0	0	10	25	289,337,364	675,145	25
Other citrus hybrids	FC	ex080520	TNE	15,000	0	0	2	35	205,059,674	288,223	2
Lemons	FC	08055010	TNE	10,000	0	6	6	51	77,884,166	143,369	70
Table grapes, fresh (21.7 to 31.10)	FC	ex08061019,08061010	TNE	1500	0	0	9	18	562,602,727	359,414	100
Apples, fresh (1.4 to 31.7)	FC	08081099,08081020,08	TNE	600	0	0	0	49	619,896,088	703,823	100
Pears, fresh, other than...in bulk (1.8 to	FC	ex08082050	TNE	1000	0	0	5	53	222,768,332	292,229	100
Apricots, fresh (1.6 to 31.7)	FC	ex08091000	TNE	3000	0	0	10	20	12,875,791	7946	0
Cherries, fresh, other than sour cherries	FC	ex08092095	TNE	800	0	0	4	12	20,485,768	6526	100
Manioc starch (manufacture of food prep)	LD	ex11081400	TNE	8000	0		68	68			13
Manioc starch (manufacture of medicame)	LD	ex11081400	TNE	10,000	0	64	68	68	4,122,413	13,126	100
Mushrooms, prepared or preserved	HI	20031020, 20031030	TNE	62,660	1	257	19	186	59,908,392	57,315	38
Potatoes, fresh or chilled: new (1.1 to 15)	FC	07019051	TNE	4000	0		3	10			100
Carrot and turnips, fresh or chilled	FC	07061000	TNE	1200	0	14	7	14	5,419,918	10,989	100
Other vegetables...sweet peppers	FC	07096010	TNE	500	0	7	2	7	89,441,947	87,277	100

(contd)

1	2	3	4	5	6	7	8	9	10	11	12
							Bound Tariffs		Total Imports		
Description	REG	HSNo	Units	Quota	Specific (1) × Global (0)	Trade Weighted applied tariffs (per cent)	In-Quota (TI) (per cent)	Out-of Quota (T2) (per cent)	Value	Quantity	Fill Rate Truncate (per cent)
Dried onions	FC	07122000	TNE	12,000	0	13	10	13	56,752,209	29,207	81
Almonds, other than bitter	FC	ex080211, ex080212	TNE	90,000	0	4	2	5	368,517,190	136,113	100
Orange juice	FC	ex20091199	TNE	1500	0	15	13	15	78,160,331	91,620	6
Grape juice (including grape must)	LD	20089943	TNE	14,000	0	4	35	95	3,449,550	3965	20

Notes:
- Description: detailed description of each TRQ tariff line from WTO in column 1.
- REG: Administration Methods in column 2.

Code	Administration Methods
AT	Applied Tariffs
FC	First-come first-served
LD	Licences on Demand
HI	Historical Importers
ST	State Trading and Producer Groups
MX	Mixed
NS	Non-specified

- HSNO: Harmonized System (HS) Numbers, typically 4–6–10 digits long in column 3. Most countries recognize the same first 6 'harmonized' digits.
- Units: the unit reported in the Notifications or TN/AG/S/5 paper in column 4.
- Quota: the quota as shown in Notification and TN/AG/S/5 paper in column 5.
- Specific versus global: allocation of quotas to a specific country or to all countries in column 6. This information is available in the Notifications.
- Trade weighted applied tariffs: weighted average applied out-of-quota tariffs in column 7. It refers to the ratios of applied tariff revenue over total value import by TRQ.
- Bound tariffs T1 and T2: columns 8 and 9 indicate the in-quota and out-of quota bound tariffs. These data come from ERS/USDA data base, which is available at: *http://www.ers.usda.gov/db/Wto/WtoTariff_database/*
- Total value imports: total value import reported in IDB in column 10. FAOSTAT were used to check some values imports.
- Total quantity import: total quantity import reported in IDB in column 11. Total in-quota and out-of-quota imports but not non-quota.
- Fill rate truncated: the simple average fill rate (defined as the average of the ratios of the value of in-quota imports over the value of quota) except it takes a maximum value of 100 per cent in column 12. It ignores over-quota imports.

Source: Database developed by de Gorter and Kliauga (2005).

TABLE 5.17: Mangoes: NTMs Faced by a Sample of Indian Firms in Export Markets

Export Market	NTMs
USA	Packaging and labelling regulations on mango pulp, import ban on fresh mangoes. Uniform size and adherence to the FDA guidelines on fruits and vegetables
Japan	The packaging and labelling regulations for mango pulp exports, import ban on fresh mango export
Europe	Uniform size of fresh mangoes, aflatoxin residue limits in dried mangoes, excessive documentation, pre- and post-harvest testing and other procedural requirements related to use of certain pesticides and fungicides, labour standards (child labour issue)
UK	Pesticide residue level, uniformity in ripeness, colours, and size for fresh mangoes, pesticide residue regulations regarding mango chutneys and pickles
Germany	Pesticide residue regulations in mango chutneys and pickles
Spain	Pesticide residue regulations in mango chutneys and pickles
Gulf countries	Uniform size of fresh mangoes, ban on export of Neelam mangoes, legalization of documents
	Uniformity in ripeness and size, checks at port of entry by non-technical personals with vague and arbitrary specification
Kuwait	Uniformity in ripeness and size

Source: Wignaraja et al. (2001).

which also happens to be the market where Indian exporters face least competition.

NOTES

1. This Chapter draws upon Diop and Jaffee (2004).
2. The AMS was defined in the Uruguay Round Agreement on Agriculture (URAA) as an aggregate measure, designed to quantify and compare countries' annual levels of domestic support. In principle, the AMS aggregates the effects of *all* trade-distorting policies (direct subsidies plus implicit subsidies from border measures) into a single measure of support. But the WTO definition of AMS does not explicitly take into account protection granted through tariffs. The impact of tariffs is, however, implicitly included insofar as they lead to, or make it possible to sustain, a domestic price above world prices. In fact, price support measures have been the most important type of policy measure within the category of assistance that is not exempt from Uruguay Round reduction commitments. Price support can be provided either through administered prices (involving transfers from consumers) or through certain types of direct payments from governments. For the purpose of total AMS calculations, price support is generally measured by multiplying the gap between the applied administered price and a specified fixed external reference price ('world market price') by the quantity of production eligible to receive the administered price. Calculation details are specified in Annexes 3 and 4 of the Agreement on Agriculture.
3. Where it is not practicable to calculate a product-specific AMS as set out in the Agreement on Agriculture, provisions are made of an 'Equivalent Measurement of Support' (EMS). The EMS is generally calculated on the basis of budgetary outlays—the money spent by governments to support a product, for example, rather than market price support calculated with respect to a fixed external reference price.

4. A recent World Bank mission focusing specifically on standards issues in horticulture has identified a number of similar problems in other commodities as well. This Report is expected to be complete in 2005–06.
5. For example, adherence to maximum residue levels (MRLs) of pesticides in food and the possession of a phytosanitary certificate are legal requirements for any exporter. Many other requirements such as traceability, adherence to good agricultural practice (GAP), and the possession of a hazard analysis and critical control point (HACCP) system are not legally mandated but may be imposed by the buyer. Reardon et al. (2002) note that all export firms must now have sophisticated quality assurance systems. Increasingly, retailers have expanded standards to include broader social, environmental and ethical issues (Dolan and Humphrey 2000).

6 Priorities for Domestic Reform and International Engagement

Looking ahead, India needs to simultaneously lower the logistical tax at home and take an aggressive position in the WTO negotiations. The inefficiencies identified in the services sectors supporting horticulture reflect larger problems affecting trade in all agricultural products. Thus greater liberalization of services supporting agriculture, including air, maritime, and road transport, as well as improvements in storage and marketing infrastructure and services are critical if India is to emerge as an important exporter in agriculture in general and in horticulture in particular. At the same time, it is essential that India plays a more active role in the WTO negotiations, seeking not just significantly lower levels of protection in destination markets, but also much greater transparency, simplicity, and predictability of foreign trade regimes.

There already exist a number of studies/reports that have persuasively argued for further liberalization of services sectors both at home and abroad to stimulate and sustain India's high growth rate. The World Bank's report 'Sustaining India's Services Revolution: Access to Foreign Markets, Domestic Reforms and International Negotiations' is one such report that contains a detailed discussion of the priorities and strategies for reform in the services sector. Two World Bank reports are under preparation that will provide comprehensive policy recommendations on trade facilitation issues and on standards in horticulture. These reports, as well as other work by the Government of India, the World Bank, and other research, suggest that the following measures could support trade in Indian agriculture.

WTO NEGOTIATIONS: TARIFFS AND STANDARDS

India should use the negotiations under the Doha Development Agenda to eliminate distortionary protection at home and abroad. In particular, it should address the opacity and variability of tariffs abroad, as well as the escalation in foreign tariffs that dampens the incentive to move into more highly processed categories. At the same time, it should consider elimination of high domestic tariffs which tax consumers far from production regions and in the off-season and end up subsidizing not the farmer, but the inefficient domestic logistics and storage systems.

There is a need to distinguish dispassionately between the SPS requirements imposed abroad that are unnecessarily burdensome and protectionist, from measures that serve legitimate safety and health concerns. The former should be the basis for bilateral consultations and, where necessary, should be challenged under the WTO's Agreement on SPS standards. Accepting the critical role of standards in horticulture trade, given the evidence that private standards are becoming more important than mandatory standards, India should adopt a coherent, forward-looking strategy to position its horticulture in regional and international markets. In particular, there is a need to strengthen and build confidence in India's food safety and agricultural health management systems—including enforcement of pesticide-related regulations, assurance of proper hygienic conditions in factories, and enhanced credibility of phyto-sanitary certifications.

AIR TRANSPORT INFRASTRUCTURE AND SERVICES

The Committee on Civil Aviation Sector in its report has suggested a number of measures to reduce air transport costs and improve the quality of services including: (i) substantial lowering of excise duty and sales tax on ATF and freedom to source ATF from both public and private sector oil companies; lowering of airport charges and liberal fiscal regime; (ii) foreign equity investment norms pertaining to both domestic and international scheduled air transport services should be further liberalized to allow from 49 to 100 per cent foreign investment in different air transport services and allowing foreign airlines to invest in airline sector; and (iii) gradual privatization of airports, starting with Mumbai and Delhi, which has been recently completed.

MARITIME TRANSPORT INFRASTRUCTURE AND SERVICES

India's TFYP report contains the following recommendations to improve the quality of port services and to reduce maritime costs: (i) Amendment of the Major Port Trusts Act, 1963, to enable faster corporatization of major ports staring with JNPT and Haldia (this Bill has already been introduced in Parliament in 2001); (ii) development of multi-modal transport from the two gateway ports, JNPT and Chennai, which are capable of berthing mother vessels, by a rail/road bridge and equipping them with modern container handling facilities; (iii) revision of the tariff policy with the overall objective of moving towards competitive pricing, elimination of cross-subsidization, and improvement in standards of services. For this to happen, the Tariff Authority of Major Ports (TAMP) should be restructured as an appellate body; and (iv) steps should be taken to involve private sector in a big way to augment port capacity and to execute projects on a BOT basis.

POLICIES TO CREATE A COMMON DOMESTIC MARKET AND TO INCREASE COMPETITION AT HOME

A recent World Bank report on Indian agriculture makes the following policy suggestions to increase domestic competition in agricultural marketing and storage activities: (i) remove permanently all storage, marketing, and credit restrictions on all commodities except during periods of emergencies; (ii) remove all reservations of agro-industrial activities for SSI;

(iii) The government should encourage states to amend the APMA to allow the non-governmental sector to develop and operate agricultural wholesale markets and farmers to market their produce wholesale outside the state-regulated markets (already being done in some states); and

(iv) The government can foster the development of contractual arrangements in agriculture by facilitating the creation of producer organizations, legislating appropriate contract law and enforcing it effectively.

Appendix I

ESTIMATION OF NPCs

The study relies on price comparison-based estimates of the nominal protection of India's principal horticultural crops. The crops covered are apples, bananas, grapes, mangoes, and oranges among fruits; and green peas, okra, onions, potatoes, and tomatoes among vegetables. The NPC at the commodity level for selected fruits and vegetables are estimated at four major consuming centres in India located in the cities of Delhi, Mumbai, Kolkata, and Chennai. These cities are located in the northern, western, eastern, and southern regions of the country respectively, and are assumed to be representative of markets in these regions. These cities account for a majority of trade in fruits and vegetables.

The price competitiveness of Indian commodities is measured through the NPC. This coefficient is defined as the ratio of domestic prices in a market to the international reference price. An NPC that is greater than unity implies that imports have a price advantage over domestic produce which renders domestic produce uncompetitive. It is, however, important to remember that there are several dimensions along which horticultural commodities compete with each other. Other than price, quality aspects such as freshness, size, colour, and taste are also important. However, the NPC estimate does not capture these.

Domestic Price: In estimating the NPC, domestic regional price is taken as the average monthly wholesale price prevailing in the local mandis (produce markets) in these cities. These are obtained from the Indian Horticultural Database for various years, published by the National Horticulture Board (NHB). Yearly prices are obtained as the weighted average of monthly prices, weighted by monthly arrivals.

Border Price: The procedure followed for estimating international prices depends on whether the commodity in question is actually being imported into the country or not.

For products which are being imported into India, month-wise information on quantity of imports, and their rupee value was obtained from the office of the Directorate General of Commercial Intelligence and Statistics (DGCIS). International import prices are estimated as the unit import price (CIF) which is derived by dividing the value of imports by quantity. Using this methodology allows us to get a rich database of the spatial variation in price competitiveness of produce across different regional markets and also of the time variations in

NPCs across months. This is important given the seasonality of production.

For products which are potential importables, but for which there have been no actual imports, the comparison is with made with prices of potential low-cost exporters in India's neighbourhood. Table A1.1 provides a list of these assumed competitors. Unit export prices (FOB) are calculated using information on actual export quantities and export earnings obtained from the FAO database. The reference CIF import price is estimated by adding international freight charges from these countries to Mumbai and is based on information obtained from actual transporters (data pertain to 2004). The freight charges have been deflated to cover the study period using the liner freight indices published by the UNCTAD Secretariat in various issues of the Review of Maritime Transport. Costs quoted in dollars were converted to rupee terms by multiplying with the yearly exchange rate.

Port Clearance and Other Charges: These estimates have been obtained from importers and traders' surveyed for the study.

Transportation Costs: Since most of the trade in fresh produce takes place in Mumbai, it is assumed that all imports arrive here first and are then transported to the various centres by road. Transportation charges were estimated by contacting various transport companies and obtaining their rates.

Table A1.1: Assumed Potential Competitors

Product	Assumed competitor
Banana	Malaysia
Mango	Pakistan
Tomato	China
Okra	Malaysia
Onion	Egypt

Importers' Margin: Traders' margin is estimated at 6 per cent of the landed cost (import CIF price + port charges).

International Reference Price: The international reference price is the border price adjusted for port charges, transportation costs, and importers' margin. The NPC is calculated as the annual international reference price divided by the domestic price. All-India aggregates are obtained by averaging regional NPCs, using yearly arrivals at each centre as weights.

Appendix 2

SURVEY DESIGN

The study is based on a primary survey that covered 1400 farmers, 200 commission agents, and 65 traders across 15 Indian states for 10 commodities.

Choice of commodities: There are three criteria that have been used to prioritize among commodities: (i) share of that crop's production in total volume of production within its broad category, (ii) share of total production by value; and (iii) its current importance in trade, since this can also be taken as a rough measure of its future importance. On this basis we arrive at a basket comprising five fruits, and an equal number of vegetables (Table A2.1). This basket covers the most important crops in the fruits and vegetables categories: in fruits this

Figure A2.1: Composition of Farmers Surveyed Commodity-wise

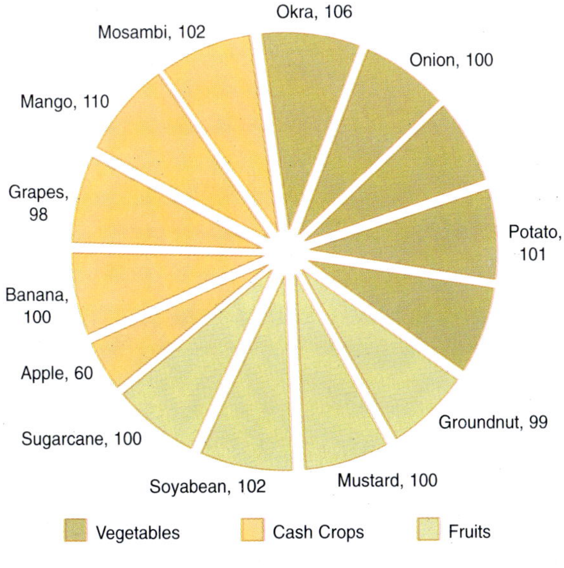

TABLE A2.1: Commodity-wise Production in the States Surveyed

Commodities	Production in the selected states as a % of total production	% of total production of fruits/ vegetables in India	% of total production by value
Apple	99	2.7	7.9
Banana	70	33	19.6
Grapes	90	2.8	7.1
Mango	58	23.3	53.8
Mosambi	92	2.8	3.5
Average	78	64.7	91.9
Green peas	77	2.3	8.5
Okra	81	3.8	9.5
Onion	56	5.9	5.3
Potato	75	27.6	35.2
Tomato	46	8.4	12
Average	63	53.7	70.6

Source: Indian Horticultural database 2003; NHB Own Calculations.

covers about 65 per cent of total production by volume and 92 per cent by total value. Comparable figures for vegetables are 54 per cent and 71 per cent (Table A2.1).

For the farmers' survey, a stratified random sampling technique was followed. For each commodity up to five major producing states were chosen. In each state, one or two districts were selected, and in these districts, farmers from different villages were chosen randomly according to the following criteria.

Choice of states: From a trade point of view the objectives while choosing the states to be covered for each crop are: (a) to cover states in different socio-economic conditions, and (b) to cover as large a range of cost differentials as possible. Since the second is not directly observable, we use productivity as a proxy.

Thus two different criteria were used to create a matrix of states: (i) level of per capita income (which served as a proxy for the diverse socio-economic conditions that prevail across states), and (ii) crop yield or productivity. The following methodology was used to choose the states.

- Choose two high income states, one with the highest productivity for that crop in that income category, and one with the lowest;
- Choose two low income states using the same criterion;
- The fifth state is chosen as one with the highest volume of production.

Figure A2.2: States Surveyed

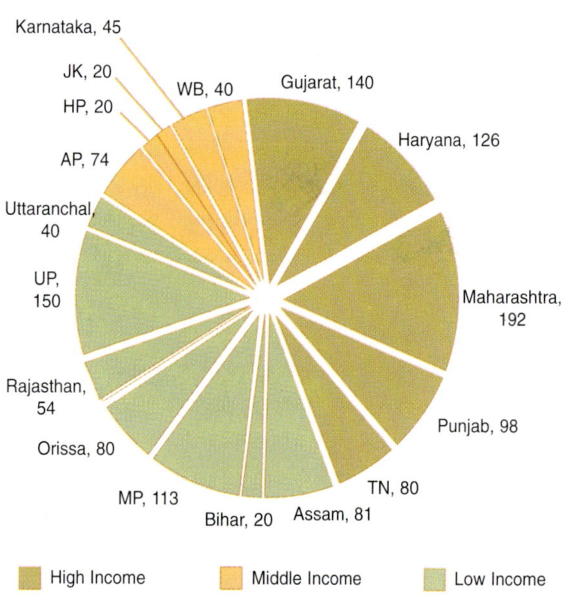

District: In each state, one or two districts with the highest production for the commodity under study were chosen. This selection was based on production data available with the state statistical abstract.

The final choice of commodities, states, and the districts is listed in Table A2.3.

Block/village: In each district, two blocks with the highest production were chosen. In each block one village was covered.

Farmers: In each village, at least five farmers have been surveyed, ensuring that at least one farmer belongs to each of the four land holding classes (large, medium, small, and marginal). The sample size for farmers is about 20 per state and 100 per commodity. In order to maintain the randomness of the sample the following method was followed:

a. Identify the total number of households (HH) in the village producing the commodity under survey;

b. Identify the sampling interval (ratio of total HH and five);

c. Generate one random number between 1-sampling interval (by lottery);

d. Pick this HH as the first sample;

e. Subsequently using right hand rule identify the second household by adding the sampling interval;

TABLE A2.2 Final Choice of Commodities, States, and Districts

Commodities	States	Districts
Apple	Jammu and Kashmir	Srinagar, Anantnag
	Himachal Pradesh	Shimla, Kulu
	Uttaranchal	Nainital
Banana	Maharashtra	Jalgaon, Nanded
	Tamil Nadu	Tiruchirapalli
	Gujarat	Anand, Surat
	Madhya Pradesh	Khandwa, Barwani
	Assam	Carpeta, Nagaon
Grapes	Punjab	Bhantida, Phirozpur
	Haryana	Fatehabad, Sirsa
	Maharashtra	Nasik, Sangali
	Karnataka	Belgaum, Bangalore
Mango	Maharashtra	Ratnagiri, Sindhugarh
	Uttar Pradesh	Lucknow
	Gujarat	Junagadh, Valsad
	Andhra Pradesh	Krishna, Chittoor
	Orissa	Angul, Ganjam
Mosambi	Maharashtra	Jalna, Nagpur
	Punjab	Ferozepur, Hoshiarpur
	Andhra Pradesh	Anantapur, Nellore
Okra	Haryana	Karnal
	Gujarat	Surat, Baroda
	West Bengal	Nadia, 24 Parganas
	Bihar	Vaishali, Patna
	Orissa	Jajpur
Onion	Maharashtra	Nasik, Pune
	Gujarat	Bhavnagar, Junagadh
	Tamil Nadu	Coimbatore
	Uttar Pradesh	Bulendshekhar, Ghazipur
	Orissa	Angul, Kalahandi
Peas	Punjab	Amritsar, Hoshiarpur
	Haryana	Sonipat, Kurukshetra
	Uttar Pradesh	Jalaun, Jhansi
	Uttaranchal	Udhamsinghnagar, Nainital
	Madhya Pradesh	Ujjain, Jabalpur
Potato	Gujarat	Banaskantha, Sabarkantha
	Haryana	Kurukshetra, Karnal
	West Bengal	Hoogly, Burdwan
	Uttar Pradesh	Agra
	Assam	Barpeta, Nagaon
Tomato	Maharashtra	Pune, Nasik
	Gujarat	Kheda, Mehsana
	Karnataka	Haveri
	Orissa	Ganjam, Keonjhar
	Assam	Nagaon, Sonitur

f. In either case if the sample HH happens to be landless or does not cultivate the crop under study, move to the next HH till the HH with the crop under study has been found.

Wholesalers/Commission Agents: The sample size for commission agents is 15 per commodity and 3–4 per state. One commission agent was randomly (by lottery) chosen from the following mandis (regulated markets) in a district: the district mandi and two village mandis (nearest to the two villages being surveyed).

Traders: A total number of 65 traders involved in the export and import of fresh fruits and vegetables have been interviewed for the study. The cities of Mumbai, Hyderabad, Chennai, Bangalore, Kolkata, Maldah, and Nasik have been covered for the survey of exporters and importers.

Bibliography

Aksoy, M. Ataman and John C. Beghin (eds) (2004), *Global Agricultural Trade and Developing Countries*, World Bank, Washington, DC.

Bannerjee, G.D. (2005), 'Poised for a Golden Revolution', *Times Agriculture Journal*, 8 June.

Balat, Jorge, F. and Guido Porto (2005), 'Globalization and Complementary Policies: Poverty Impacts in Rural Zambia'. Unpublished Working Paper, World Bank, Washington, DC.

Benavene, Jose Miguel (2004), 'Technological change in Chile's grape and wine producing industry', Department of Economics, University of Chile, December.

BIG's Easy Reference Customs Tariff 2005–6 (2005), 24th budget edition, Academy of Business Studies, New Delhi.

Calvin, Linda, Roberta Cook et al. (2001), 'US Fresh Fruit and Vegetable Marketing: Emerging Trade Practices, Trends and Issues', United States Department of Agriculture, Agricultural Economic Report No. 795, January.

Centre d'Etudes Prospectives et d'Informations Internationales (CEPII), *www.cepii.fr/anglaisgraph/bdd/distances.htm*

Chand, Ramesh (2002), *Trade Liberalisation, WTO and Indian Agriculture—Experience and Prospects*, Mittal Publications, New Delhi.

—— (2004), 'Agricultural Growth during the Reforms and Liberalisation: Issues and Concerns', Policy Brief 20, National Centre for Agricultural Economics and Policy Research.

CII (Confederation of Indian Industry) and McKinsey (1997), *Modernising the Indian food Chain*.

Clark, Ximena, David Dollar, and Alejandro Micco (2002), 'Maritime Transport Costs and Port Efficiency', World Bank Policy Research Working Paper 2781, February.

Collinson, Chris (2001), 'The Business Costs of Ethical Supply Chain Management: Kenyan Flower Industry Case Study', NRI Report No. 2607, Natural Resources Institute, May.

Cook, Roberta L. (2004), 'Supermarket Challenges and Opportunities for Fresh Fruit and Vegetable Producers and Shippers: Lessons from the US Experience', Paper presented at the Conference on Supermarkets and Agricultural Development in China—Opportunities and Challenges, Shanghai, China, 24 May.

Deininger, Dina Umali (2004), 'India Re-energizing the Agricultural Sector to Sustain Growth and Reduce Poverty', Report No. 27889-IN, World Bank, 30 July.

Department of Agriculture, Fisheries and Forestry (2004), Australian Government, *Fact Sheet for Australian Horticulture*, June.

De Gortue, H. and E. Kliauga (2005), 'Reducing Tariffs versus Expanding Tariff Rate Quotas' in *Agricultural Trade Reform and the Dorha Development Agenda*, K. Anderson and W. Martin (eds), Palgrave Macmillan and the World Bank.

Dimitri, Carolyn, Abebayehu Tegene, and Phil R. Kaufman (2003), 'US Fresh Produce Markets',

United States Department of Agriculture, Agricultural Economic Report No. 825, September.

Diop, Ndiame and Steven M. Jaffee (2004), 'Fruits and Vegetables: Global Trade and Competition in Fresh and Processed Product Markets', in M. Ataman Aksoy and John C. Beghin (eds), *Global Agricultural Trade and Developing Countries*, World Bank, Washington DC.

Directorate General of Commercial Intelligence and Statistics (DGCIS), database, 1994–2003.

Directorate of Economics and Statistics (2003), *Report of the Committee on Capital Formation in Agriculture*, Department of Agriculture and Co-operation, Ministry of Agriculture, Government of India, March.

Directorate of Marketing and Inspection (2002), *Agricultural Marketing*, A National Level Quarterly Journal on Agricultural Marketing, Vol. XLV, No. 2, July–September. Department of Agriculture and Cooperation, Ministry of Agriculture, Government of India.

Export Import Bank of India (2002), *Potential for Exports of Horticulture Products from Bihar and Jharkhand*, Occasional paper No. 94, Quest Publications, December.

Food and Agriculture Organization (2004), http://faostat.fao.org/faostat/collections?subset=agriculture 1990–2004.

Gandhi, Vasant P. and N.V. Namboodiri (2002), *Fruit and Vegetable Marketing and its Efficiency in India: A Study of Wholesale Markets in the Ahmedabad Area*, Centre for Management in Agriculture, Indian Institute of Management, Ahmedabad, June.

____ (forthcoming), *Marketing of Fruits and Vegetables in India: A Study Covering the Ahmedabad, Chennai and Kolkata Markets*, Indian Institute of Management, Ahmedabad.

Government of India (2004), *Report on Working Group on Trade facilitation*, October.

Gulati, Ashok, Anil Sharma, Kailash Sharma, Shirpa Das, and Vandana Chhabra (1994), 'Export Competitiveness of Selected Agricultural Commodities', NCAER Working Paper.

Gulati, Ashok, Garry Pursell, and Kathleen Mullen (2003), *Indian Agriculture since the Reform: Performance, Policy Environment and Incentives*, IFPRI, May.

Gulati, Ashok, James Hanson, and Garry Pursell (1990), *Effective Incentives in India's Agriculture: Cotton, Groundnuts, Wheat and Rice*, WPS 332, World Bank, January.

Gulati, Ashok and Sudha Narayanan (eds) (2003), *The Subsidy Syndrome in Indian Agriculture*, Oxford University Press, New Delhi.

Huang, Sophia Wu (2004), 'Global Trade Patterns in Fruits and Vegetables', USDA Agriculture and Trade Report Number WRS-04–06, Washington, DC.

Hummels, David (2001), 'Time as a Trade Barrier', Purdue University, July (mimeo).

India Exporter Guide Annual (2004), 'United States Department of Agriculture, Foreign Agricultural Service', GAIN Report No. IN4103, September.

Ingco, Merlinda D. (1997), 'Has Agricultural Trade Liberalization Improved Welfare in the Least-Developed Countries? Yes', World Bank Policy Research Working Paper 1748, April.

Ingco, Merlinda D. and John D. Nash (eds) (2004), *Agriculture and the WTO: Creating a Trading System for Development*, World Bank and Oxford University Press.

Jaffee, Steven (2003), 'From Challenge to Opportunity; Transforming Kenya's Fresh Vegetable Trade in the Context of Emerging Food Safety and Other Standards in Europe', Agriculture and Rural Development Discussion Paper 2, World Bank, December.

Jaffee, Steven and Spencer Henson (2004), 'Standards and Agro-Food Exports from Developing Countries: Rebalancing the Debate', Working Paper Series 3348, June, World Bank, Washington, DC.

Jaffee, Steven, S. Henson et al. (2005), 'Food safety and Agricultural Health Standards: Challenges and Opportunities for Developing Country Exports', World Bank Report No. 31207, January.

Jha, Veena, Sarika Gupta, James Nedumpara, and Kailas Karthikeyan (2004), *Trade Liberalisation and Poverty in India*, United Nations Conference on Trade and Development, April.

Kapila, Raj and Uma Kapila (2002), *A Decade of Economic Reforms in India*, Academic Foundation, New Delhi.

Landes, Maurice R. (2004), 'The Elephant is Jogging: New Pressures for Agricultural Reform in India', USDA Economic Research Service, Washington, DC.

Limão, Nuno and Anthony J. Venables (1999), 'Infrastructure, Geographical Disadvantage, and Transport Costs', WPS-2257, World Bank, December.

Mariental India Pvt. Ltd. (2003), 'Comprehensive Study on Processing of Apple', Report for the Ministry of Food Processing Industries, Government of India.

Martin, William J. (2003), 'Global Agricultural Market Access Liberalization under the Doha Development Agenda: Interests and Options for India', (mimeo), World Bank, August.

Ministry of Agriculture (2002), *Report of Inter-Ministerial Task Force on Agricultural Marketing Reforms*, Government of India, May.

Ministry of Civil Aviation (2003), *Report of the Committee on a Road Map for the Civil Aviation Sector*, Government of India, November.

Mullen, Kathleen, David Orden, and Ashok Gulati (2005), 'Agricultural Policies in India: Producer Support Estimates 1985–2002', MTID Discussion Paper No. 82, International Food Policy Research Institute, February.

Naik, Gopal (2004a), 'Bridging the Knowledge Gap for Building Competitive Agriculture: The Case of Grapes in India', Centre for Public Policy, Indian Institute of Management, Bangalore, November.

____ (2004b), 'Bridging the Gap in Maize Production Technology: The case of maize in India', Centre for Public Policy, Indian Institute of Management, Bangalore.

National Horticulture Board (NHB), National Horticulture Board, Indian Horticulture Database, various issues.

____ *Wholesale and Retail Marketing of Fruits and Vegetables in Metropolitan Cities (1995–1999)*, annual database brought out by National Horticulture Board, Gurgaon, Haryana.

Nayyar, Deepak and Abhijit Sen (1994), 'International Trade and the Agricultural Sector in India', *Economic and Political Weekly*, 14 May.

Panagariya, Arvind (forthcoming), 'Agricultural Liberalisation and the Least Developed Countries: Six Fallacies', World Economy: Global Trade Policy 2005 (mimeo).

Parikh, Kirit S. (2004), *Agricultural Trade Liberalization: Impact of Alternative Proposals on India*, A Study prepared for United Nations Conference on International Trade and Development, 15 June.

Pursell, Garry and Anju Gupta (1998), 'Trade Policies and Incentives in Indian Agriculture, Methodology, Background Statistics and Incentive Indicators, 1965–95, Sugar and sugarcane', Background Paper No. 1, World Bank, Washington DC.

Pursell, Garry and Ashok Gulatin (1993), *Liberalizing Indian Agriculture: An Agenda for Reform*, World Bank, Policy, Planning, and Research Working Paper 1172.

Radelet, Steven and Jeffrey Sachs (1998), 'Shipping Costs, Manufactured Exports and Economic Growth', Earth Institute, Columbia University, 1 January.

Rao, C.H. Hanumantha (2001), 'WTO and Viability of Indian Agriculture', *Economic and Political Weekly*, Vol. 36, 8 September.

Raghunath, S. and D. Ashok (2004), 'Delivering Simultaneous Benefits to the Farmer and the Common Man: Time to Unshackle the Agricultural Produce Distribution System', Indian Institute of Management, Bangalore, June.

Reardon, T. and J.A. Berdegue (2003), 'The Rapid Rise of Supermarkets in Latin America: Challenges and Opportunities for Development', *Development Policy Review*, Vol. 20, No. 4, pp. 371–88.

Roberts, Donna and Barry Krissoff (2004), 'Regulatory Barriers in International Horticultural Markets', United States Department of Agriculture, WRS-04–01, January.

Roy, Jayanta (2002), 'Towards International Norms for

Indirect Taxes and Trade Facilitation in India', Background paper prepared for the Task Force on Indirect Taxes, GoI, October.

Roy, Jayanta (2005), 'Trade Facilitation and FDI: A Business Perspective', presentation made at a conference in Colombo.

Sanchez, Jorge and Sandra Kipe (2004), 'US Horticultural Exports to Asia 2004', United States Department of Agriculture, August, 2004.

Sardana, Vijay (2004), 'New Framework for WTO Negotiations and its Implications and Suggestion for Indian Food and Agriculture Sector', Policy Discussion Paper, Policy paper published by Centre for International Trade in Agriculture and Agro Based Industries, New Delhi.

Sekhar, C.S.C. (2004), 'Agricultural Price Volatility in International and Indian Markets', *Economic and Political Weekly*, 23 October.

Sibal, Vikrant (2004), *Indian Sugar Industry*, India Research, HDFC Securities, 9 September.

Singh, N.K. and Jessica S. Wallack, (2005), 'Moving India: Policies and Priorities in Transport Sector Reform', Paper prepared for the Stanford Center for International Development, Sixth Annual Conference on Indian Economic Reforms, Stanford CA, 3–4 June.

Srinivasan, P.V. (2004), 'Managing Price Volatility in an open economy environment: The case of Edible Oils and Oilseeds in India', MTID Discussion Paper No. 69, IFPRI, International Food Policy Research Institute, Washington DC.

Sawhney, Aparna (2005), *Quality Measures in Food Trade: The Indian Experience*, Blackwell Publishing Ltd., Malden, MA.

Swarup, R. and B.K. Sikka (1987), *Production and Marketing of Apples*, Mittal Publications, Delhi.

United Nations Conference on Trade and Development (2004), *Review of Maritime Transport 2004*, Report by the UNCTAD Secretariat, United Nations, New York and Geneva.

Valdes, Alberto and William Foster (2003), 'Special Safeguards for Developing Country Agriculture in WTO Negotiations', Paper presented at the International Conference Agricultural Policy Reform and the WTO: Where Are We Heading? Capri (Italy), 23–26 June.

Virmani, Arvind (2003), 'Customs Tariff Reform', ICRIER Policy Brief, January.

Whitaker, Meri and Shashi Kolavalli (2004), *Floriculture in Kenya*, Uwazo-Creative Directions.

Wignaraja, G., B. Debroy, P.D. Kaushik, and Fred Nixon (2001), 'Non-Tariff Barriers to Indian Exports', Commonwealth Secretariat Publication, United Kingdom.

World Bank (1997), 'The Indian Oilseed Complex: Capturing Market Opportunities', Vols I and II, Report No. 15677-IN, 31 July.

—— (1999), 'Indian Foodgrain Marketing Policies: Reforming to Meet Food Security Needs', Vols I and II, Report No. 18329-IN, 17 August.

—— (2001), 'Forging Subregional Links in Transportation and Logistics in South Asia', Uma Subramanian and John Arnold, January.

—— (2002), 'Global Economic Prospects and the Developing Countries, Making Trade Work for the World's Poor', Report No. 23590, January.

—— (2004), 'India: Reenergizing the Agricultural Sector to Sustain Growth and Poverty Reduction', Rural Development Unit, South Asia Region, July.

—— (2005a), Bangladesh: Growth and Export Competitiveness, Poverty Reduction and Economic Management Sector Unit, May.

—— (2005b), 'Food Safety and Agricultural Health Standards: Challenges and Opportunities for Developing Country Exports', Poverty Reduction and Economic Management Trade Unit and Agriculture and Rural Development Department, Report No. 31207, 10 January, Washington, DC.